I0047703

Confocal Microscopy

Confocal Microscopy

Jian Liu and Jiubin Tan
Harbin Institute of Technology (HIT)

Morgan & Claypool Publishers

Copyright © 2016 Morgan & Claypool Publishers

All rights reserved. No part of this publication may be reproduced, stored in a retrieval system or transmitted in any form or by any means, electronic, mechanical, photocopying, recording or otherwise, without the prior permission of the publisher, or as expressly permitted by law or under terms agreed with the appropriate rights organization. Multiple copying is permitted in accordance with the terms of licences issued by the Copyright Licensing Agency, the Copyright Clearance Centre and other reproduction rights organisations.

Rights & Permissions
To obtain permission to re-use copyrighted material from Morgan & Claypool Publishers, please contact info@morganclaypool.com.

ISBN   978-1-6817-4337-0 (ebook)
ISBN   978-1-6817-4336-3 (print)
ISBN   978-1-6817-4339-4 (mobi)

DOI   10.1088/978-1-6817-4337-0

Version: 20161201

IOP Concise Physics
ISSN 2053-2571 (online)
ISSN 2054-7307 (print)

A Morgan & Claypool publication as part of IOP Concise Physics
Published by Morgan & Claypool Publishers, 40 Oak Drive, San Rafael, CA, 94903 USA

IOP Publishing, Temple Circus, Temple Way, Bristol BS1 6HG, UK

Contents

Preface   vii

Acknowledgments   viii

Author biography   ix

1   Confocal microscopy and its application in China   **1-1**

1.1   A brief review of confocal microscopy   1-1
1.2   Resolution   1-2
1.3   Standardization in China   1-3
     References   1-4

2   Point spread function model   **2-1**

2.1   Lens imaging   2-1
2.2   PSF in confocal imaging   2-5
     References   2-9

3   Incoherent three-dimensional optical transfer function   **3-1**

3.1   Development of 3D optical transfer function   3-1
3.2   3D imaging models of CM   3-2
3.3   3D-OTF of CM   3-4
3.4   3D-OTF of differential CM   3-6
     References   3-10

4   Decoupling criteria for three-dimensional optical microscopic measurement   **4-1**

4.1   Introduction   4-1
4.2   Decoupling model for the measurement of thin samples   4-2
4.3   Decoupling model for the measurement of a deep groove sample   4-4
4.4   Experiments   4-6
     References   4-7

5   Pupil filter design   **5-1**

5.1   Phase rotation transformation   5-2
5.2   The design method of filters with global minimizing side lobes   5-7
     References   5-10

6   Confocal axial peak extraction algorithm   **6-1**

6.1   Introduction   6-1
6.2   Centroid method for localization of confocal peak   6-2
6.3   Nonlinear fitting method for peak localization   6-2
6.4   Deviation analysis for localization of confocal axial peak   6-6
      References   6-9

7   Differential confocal microscopy   **7-1**

7.1   Introduction   7-1
7.2   Application of DCM in China   7-2
7.3   The Basic principle of DDCM   7-3
      References   7-7

8   Medium aided scattering measurement   **8-1**

8.1   Introduction   8-1
8.2   The principle of medium aided scattering confocal microscopy   8-1
8.3   Analysis of deposition uniformity of a fluorescent medium layer 8-2
8.4   Error analysis and height correction of the medium layer   8-4
8.5   Application of medium aided scattering confocal microscopy   8-4
      References   8-6

9   Scanning technology   **9-1**

9.1   Introduction   9-1
9.2   Scanners   9-1
9.3   Raster scanning   9-2
9.4   α-β circular scanning   9-3
      References   9-5

10   Confocal profilometer   **10-1**

10.1 Introduction   10-1
10.2 Basic principle   10-1
10.3 The extraction method of discrete surface   10-2
      10.3.1 Method of planar substrate conversion   10-2
      10.3.2 Method of polynomial regression   10-3
10.4 Application of confocal profilometer   10-3
      References   10-6

Preface

Many books on confocal microscopy introduced imaging theories and experimental techniques in the field of life science. Unfortunately, there are few books on confocal microscopy that elaborate on aspects of industrial metrology despite confocal microscope being an essential instrument both in life science and advanced fabrications, and its earlier applications which were measurements on integrated circuit.

In term of a 3D imaging tool, confocal microscope can be appropriate for either imaging cells or the measurement of industrial artefacts. However, junior researchers and instrument users sometimes misuse imaging concepts and metrological characteristics, such as position resolution in industrial metrology and scale resolution in bio-imaging. And, metrological characteristics or influence factors in 3D measurement such as height assessment error caused by 3D coupling effect are so far not yet identified. In this book, the authors record their practices by the working experiences on standardization and system design.

This book assumes little previous knowledge of optics, but rich experience in engineering of industrial measurements, in particular with profile metrology or areal surface topography will be very helpful to understand the theoretical concerns and value of the technological advances. It should be useful for graduate students or researchers as extended reading material, as well as microscope users alongside their handbook.

We thank Professor Tony Wilson at the University of Oxford, UK, for discussions on optical theory, and Professor Richard Leach at the University of Nottingham for discussions on metrology, and the past group members who encouraged this research.

J Liu
Harbin, China
November, 2016

Acknowledgments

This work is funded by the National Natural Science Foundation of China (Grant Nos 50905048, 51275121), Ministry of Science and Technology (Grant No. 2011YQ040087), and Defence Fundamental Researches (2013XE0873), China.

Author biography

Jian Liu

Jian Liu is Professor and vice dean of the School of Electrical Engineering and Automation, Harbin Institute of Technology, China, Honorary Professor at the University of Nottingham, UK. His academic interests lie in the theories and implementations of optical microscopes, in particular the development of confocal microscopes, applied optics and optical metrology. He is also council member of China Optical Society for Engineering and China Instrument and Control Society, and also a member of ISO/TC213 and China SAC/TC240, both for geometrical products specifications, and a board member of Journal of Microscopy, Surface Topography: Metrology & Properties and Optics Communications.

IOP Concise Physics

Confocal Microscopy

Jian Liu, He Zhang and Jiubin Tan

Chapter 1

Confocal microscopy and its application in China

1.1 A brief review of confocal microscopy

The birth of confocal microscopy (CM) can be traced back to 1957 [1]. The original concept of confocal microscopy was first proposed by Marvin Minsky at Harvard University. Minsky recalls the idea was first propounded in 1955 and the initial motivation was to reduce imaging blur and enhance contrast by reducing the size of the illuminator and detector down to a point. Certainly, the imaging field of view gets smaller when a point mask is used, however it has been proved that this can be compensated by mechanical scanning or optical scanning with a satisfactorily developed galvanometer mirror, DMD and LCD etc. Scanning techniques were again stimulated by the development of CM, after a flying spot microscope was proposed in 1951. In 1961 patenting was authorized.

However, CM did not gain much public attention, despite the approved patent until C J R Sheppard and A Choudhury first demonstrated a 1.4 times improvement in CM resolution over traditional optical microscopy, and T Wilson discussed the depth discrimination and presented the sectioning ability of CM [2–4]. Henceforth, CM became widely appreciated and heralded a new branch of modern optical microscopy. Currently, CM is highly focused and very active both in industrial metrology and life science research. Compared with AFM, MFM and EFM, scanning tunneling microscopy or near-field optical microscopy, CM has a wide range of advantages, including: low cost, contactless on samples and excellent noise inhibition abilities.

In China, the earliest theoretical discussion on CM began at the end of the 1980s at the Shanghai Institute of Optics and Fine Mechanics and the Chinese Academy of Sciences. However, during the following 20 years, the development of domestic commercial products of CM failed to materialize as expected, although CM based applications were well developed and widely used in the material and life science

doi:10.1088/978-1-6817-4337-0ch1   © Morgan & Claypool Publishers 2016

fields. Harbin Institute of Technology (HIT) began investigations on CM theory around 1998. The HIT group pioneered a number of CM based techniques, such as differential confocal sensors, digital differential confocal microscopy and phase-shift confocal interferometers as well as Sinc^2 fitting algorithms, and additionally proposed a limited energy lost (LEL) decoupling criterion for three-dimensional (3D) groove metrology etc [5–6]. In 2015, the HIT group published the first Chinese monograph on CM, to demonstrate how CM techniques have been developed and applied to Chinese industry. In 2011, China's 'Major Scientific Instruments Plan' was launched and thereafter research on commercial CM products has been robustly supported. Metrological confocal microscopes and confocal profilometers were successfully developed in HIT, and a commercial Raman-confocal microscope was achieved at Beijing Institute of Technology. The two photon microscope has mainly been researched at Suzhou Institute of Biomedical Engineering and Technology, Chinese Academy of Science and additionally at Zhejiang University, Beijing University and Xi'an Institute of Optics and Precision Mechanics. The Chinese Academy of Sciences also pioneered a number of techniques in nonlinear imaging and for structured illumination microscopes.

1.2 Resolution

The applications of CM chiefly pertain to industrial measurement or biological imaging. There is often some confusion regarding resolution among junior users and researchers, because CM resolution may be in nanometers or even higher levels in some aspects of industrial measurement, while biological imaging may require hundreds of nanometers. There may be 20 or more different definitions of resolution in a dictionary of science and technology, however the real meaning of resolution will depend on its specific application. It can be better understood if the full name of the resolution (position resolution or scale resolution) is given in the description, as position resolution describes how sensitively the system can determine the specimen laterally or vertically, while scale resolution is defined as the minimum size of possibly identified details.

The resolving ability of an optical instrument is relevant to three resolutions at least: diffractive resolution, geometrical resolution and equivalent-noise resolution. Diffractive resolution is the basic restriction for optical equipment. In free space propagation, the resolved minimum distance is defined as the Airy spot, $\delta_d = 1.22\lambda F/D$, λ is wavelength, F is focal length and D is pupil diameter. For nonlinear techniques, δ_d determines the resolving ability of an optical instrument. Geometrical resolution, $\delta_g = \Delta x R/F$, is decided by CCD pixel size, Δx is pixel distance and R is imaging distance. For geometrical imaging, super resolving imaging can be realized by moving the camera in $\Delta x/N$ to collect N frame pictures and reconstructing photos using sub-pixel algorithms. Equivalent-noise resolution is a type of short quantum noise which is closely related to temperature fluctuation, stray light, the detector's electronic noise etc.

An optical system is a low-pass filter with diffractive resolution, determined by the limited diffractive pupil size. However, super-resolving techniques exceed the limits of traditional optical systems. In 1952, Toraldo Di Francia made the first attempt to achieve high resolution beyond the Airy spot, by modulating pupil function and using weak coherent illumination. Henceforth scientists began to realize the significance of source coherence. Laser technology emerged in 1960, opening a new era of coherent optics. Pupil filtering or wave front engineering have since been developed and have been hot topics in modern optics. However it is important to be aware that side lobe growth and energy loss will accompany the compression of a main lobe, when pupil function is modulated to suppress the size of the point spread function [7–9].

Can instrument resolution be truly improved by inserting a pupil filter? Brynmor J Davis gave an interesting answer in 2004 by saying that optical cut-off does not extend in this case, but high frequency information will be partially enhanced at the cost of lower transformation at low and middle frequencies. Therefore, some initial high frequency details become visible in comparison with previous images obtained in the case of a clear pupil. In a CM system, the point-like pinhole mask can make the rising side lobe apodized so that cut-off is finally improved. This is an equivalent nonlinear response in CM. Nonlinear effects might be one of the most significant trends in today's microscopes, such as STED, STORM and I^5M, indicating that microscopy has already turned into nanoscopy.

1.3 Standardization in China

The development of CM has stimulated the 3D imaging industry, whilst providing a challenge to international standardization. Metrological theory needs to catch up with instrument development, and thereafter provide valuable guidance for users and promote applications. To the best of the authors' knowledge, the first international standard may be ISO 25178-607 which describes the nominal characteristics of CM as an area non-contact instrument and is drafted by ISO/TC213 WG16; scientists/engineers from the UK, US, Germany and China etc. are collaborating on the formulation of standards. ISO (International Organization of Standards) together with IEC (International Electrotechnical Commission) and ITU (International Telecommunications Union) publish 85% of international standards. Their member bodies comprise 98% of global GDP and 97% of the population range.

In China, the first CM national standard, Geometrical Product Specifications (GPS), Metrological Characteristics and Guide to Uncertainty of Measurement for Optical Confocal Microscopes, was drafted by HIT and administrated by SAC (Standardization Administration of the People's Republic of China)/TC240. In this national standard, the terms and definitions relating to confocal microscopes are described fully, as are the basic methods regarding 3D micro-imaging uncertainty evaluations. China's standardization is motivated by the drive of industrial development and modernization. This book is a concise summary of the authors' research into CM and its applications to industrial metrology.

References

[1] Minsky M 1957 *Confocal Patent focal Scanning Microscope.* US patent, serial, (3,013,467)

[2] Sheppard C J R and Choudhury A 1977 Image formation in the scanning microscope *Opt. Acta* **24** 1051–73

[3] Sheppard C J R and Wilson T 1978 Depth of field in the scanning microscope *Opt. Lett.* **3** 115–7

[4] White J G and Amos W B 1987 Confocal microscopy comes of age *Nature* 328 183–4

[5] Zhao W Q, Tan J B and Qiu L R 2004 Bipolar absolute differential confocal approach to higher spatial resolution *Opt. Exp.* **12** 5013–21

[6] Tan J and Wang F 2002 Theoretical analysis and property study of optical focus detection based on differential confocal microscopy *Meas. Sci. Technol.* **13** 1289–93

[7] Neil M A A, Juskaitis R and Wilson T *et al* 2000 Optimized pupil-plane filters for confocal microscope point-spread function engineering *Opt. Lett.* **25** 245–7

[8] Liu J and Tan J 2008 Synthetic complex super-resolving pupil filter based on double-beam phase modulation *Appl. Opt.* **47** 3803–7

[9] Liu J and Tan J 2008 A convex objective function based design method developed for minimizing side lobe *Appl. Opt.* **47** 4061–7

IOP Concise Physics

Confocal Microscopy

He Zhang, Jian Liu, Weibo Wang and Jiubin Tan

Chapter 2

Point spread function model

In comparison with the transfer function, the point spread function (PSF) is more intuitive in demonstrating intensity change, however it is unable to indicate bandwidth level and frequency distribution [1–3]. PSF engineering is of equal importance to the Fourier theory based transfer function, and widely used in system design. Researchers may choose either the PSF or transfer function as the tool to evaluate imaging performance. In terms of an optical system, confocal microscopy (CM) is basically the same as a conventional optical microscope except for the extremely small source and detector [4, 5]. Despite differing sizes of sources and detectors, PSF always provides an effective solution for image resolution. PSF is initially modeled using scalar theory, in order to demonstrate how resolution can be improved in CM.

2.1 Lens imaging

A thin lens with a thickness distribution of parabolic type is the simplest ideal imaging element in an optical system. It is assumed that the phase, rather than the transverse coordinates of a beam of light, change immediately after it passes through a thin lens. A concise schematic diagram of a thin lens is shown in figure 2.1, wherein d_1 and d_2 are the distances from the lens to the object and the image plane, respectively.

The transmission function of the lens can be expressed as

$$t(x, y) = P(x, y) \cdot \exp\left[\frac{ik}{2f}(x^2 + y^2)\right] \tag{2.1}$$

where f is the focal length, $P(x, y)$ is the pupil function, $k = 2\pi/\lambda$ is the wave number, and a is the radius of the thin lens. The pupil function restricting the non-opaque area can be defined as

$$P(x, y) = \begin{cases} 1 & x^2 + y^2 \leqslant a^2 \\ 0 & \text{else} \end{cases}. \tag{2.2}$$

doi:10.1088/978-1-6817-4337-0ch2   © Morgan & Claypool Publishers 2016

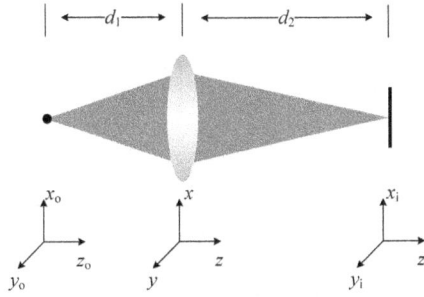

Figure 2.1. Schematic diagram of simplified thin lens imaging.

The light emitting from the point object spreads as a spherical wave, and the complex amplitude distribution prior to the lens can be given by the formula of spherical wave propagation,

$$U_{before}(x, y) = \frac{iU_0}{\lambda d_1} \exp(-ikd_1) \cdot \exp\left[\frac{-ik}{2d_1}(x^2 + y^2)\right]. \tag{2.3}$$

The amplitude immediately after the lens can be modulated using its transmission function and formulated as shown below in equation (2.4). For a clear pupil, pupil function $P(x, y)$ takes a constant value of 1,

$$\begin{aligned} U_{after}(x, y) &= P(x, y) \exp\left[\frac{ik}{2f}(x^2 + y^2)\right] \cdot U_{before}(x, y) \\ &= \frac{iU_0}{\lambda d_1} \exp(-ikd_1) \cdot P(x, y) \exp\left[\frac{ik}{2f}(x^2 + y^2)\right] \cdot \exp\left[\frac{-ik}{2d_1}(x^2 + y^2)\right]. \end{aligned} \tag{2.4}$$

The complex amplitude distribution on the detector plane is defined as $U_i(x_i, y_i)$ and the following distribution function can be obtained through the Fresnel diffraction integral,

$$\begin{aligned} U_i(x_i, y_i) &= \frac{iU_0}{\lambda d_2} \exp(-ikd_2) \iint U_{after}(x, y) \\ &\quad \times \exp\left\{\frac{-ik}{2d_2}\left[(x_i - x)^2 + (y_i - y)^2\right]\right\} dxdy \\ &= C \cdot \iint P(x, y) \exp\left[\frac{ik}{2}\left(\frac{1}{f} - \frac{1}{d_1} - \frac{1}{d_2}\right)(x^2 + y^2)\right] \\ &\quad \times \exp\left[-ik\left(x_i x + y_i y\right)\right] dxdy \end{aligned} \tag{2.5}$$

According to the imaging laws,

$$\frac{1}{f} = \frac{1}{d_1} + \frac{1}{d_2} \tag{2.6}$$

the complex amplitude on the detector can be finally given by

$$U_i(x_i, y_i) = C \cdot \iint P(x, y) \cdot \exp\left[-\frac{ik}{d_2}(x_i x + y_i y)\right] dx dy \qquad (2.7)$$

where C is a constant factor. In terms of the rotational symmetry of the pupil, the diffraction formula can be expressed in polar coordinates instead of Cartesian, in order to simplify the formula and improve computational efficiency. For an axisymmetric optical system, polar coordinates can sometimes be highly useful to researchers in analyzing energy distribution or polarization, and so the Cartesian coordinates of variables on the pupil and imaging planes can be rewritten as

$$\begin{aligned} x_i &= r_i \cos\theta \\ y_i &= r_i \sin\theta \end{aligned} \quad \text{and} \quad \begin{aligned} x &= \rho \cos\varphi \\ y &= \rho \sin\varphi \end{aligned} \qquad (2.8)$$

then

$$U_i(r_i, \theta) = K \int_0^1 \int_0^{2\pi} P(\rho, \varphi) \exp\left[\frac{1}{\lambda f} i 2\pi a \rho r_i \cos(\varphi - \theta)\right] d\varphi a \rho d\rho. \qquad (2.9)$$

By substituting the following equations of Bessel functions into equation (2.9),

$$J_0(x) = \frac{1}{2\pi} \int_0^{2\pi} \exp(ix \cos\varphi) d\varphi \qquad (2.10)$$

$$\int_0^a J_0(2\pi r r_i) 2\pi r \, dr = \pi a^2 \left[\frac{2 J_1(2\pi a r_i)}{2\pi a r_i}\right] \qquad (2.11)$$

the complex amplitude on the image plane is simplified as

$$U_i(v) = C \cdot \left[\frac{2 J_1(v)}{v}\right]. \qquad (2.12)$$

Transverse optical coordinate v can be defined as

$$v = \frac{2\pi}{\lambda} \frac{a}{d_2} r_i \qquad (2.13)$$

where $r_i = (x_i^2 + y_i^2)^{1/2}$, and r_i is the radial distance on the image plane. So the intensity response distribution of a point object on the image plane can be simplified as

$$I(v) = \left[\frac{2 J_1(v)}{v}\right]^2. \qquad (2.14)$$

$I(v)$ is the intensity point spread function (IPSF), which refers to the intensity response of a linear optical system. In an ideal thin lens, the pattern of $I(v)$ is usually called an Airy disk and it is usually used to evaluate the imaging ability of an optical system. A high resolution system has sharp main lobe and a low-intensity side lobe. For 3D microscopic imaging or measurement applications, specimen thickness cannot be neglected. Optical signals from different sections of an object are *aliasing*, and as a result the image will blur since a part of the section being imaged must have spherical aberrations. The levels of aberrations can be reflected by the size of the IPSF. A narrower IPSF is better as it is much closer to the Airy spot. Figure 2.2 shows the radial and lateral intensity distribution of $I(v)$, respectively.

An object's defocusing quantity in Cartesian coordinates can be denoted as Δz. The complex amplitude distribution on the image plane of a defocusing system is [7]

$$U_i(r_i) = \frac{i}{\lambda(f + \Delta z)} \exp\left[-ik(f + \Delta z)\right] \exp\left[-\frac{i\pi r_i^2}{\lambda(f + \Delta z)}\right]$$

$$\times \int_0^\infty P(r)\exp\left[\frac{uikr^2}{x}\left(\frac{1}{f} - \frac{1}{f + \Delta z}\right)\right] J_0\left(\frac{2\pi r r_i}{\lambda(f + \Delta z)}\right) 2\pi r\,\mathrm{d}r \tag{2.15}$$

The axial optical coordinates can be defined as

$$u = \frac{2\pi}{\lambda} \frac{a^2}{d_2^2} \Delta z. \tag{2.16}$$

So we can obtain the complex amplitude distribution near the focal plane as given by

$$U_i(v, u) = C \int_0^1 \exp\left(\frac{iu\rho^2}{2}\right) J_0(v\rho)\mathrm{d}\rho \tag{2.17}$$

where C is a constant factor. Furthermore, the 3D intensity distribution can be expressed as

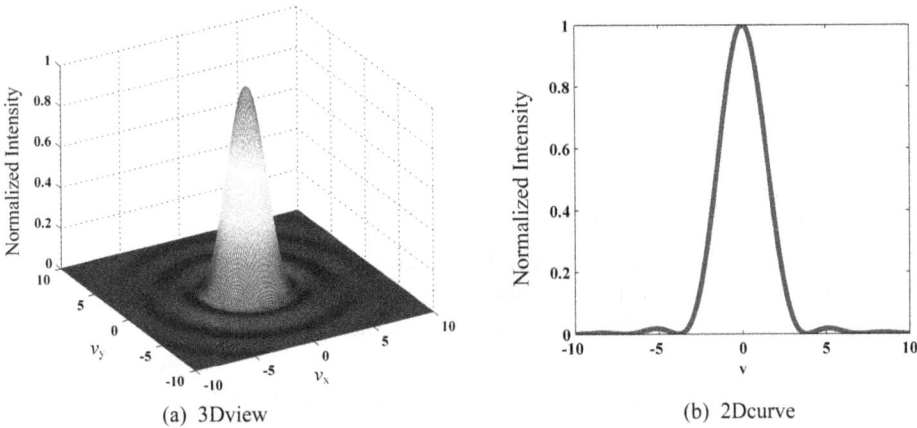

(a) 3Dview   (b) 2Dcurve

Figure 2.2. Intensity distribution of a point-like object.

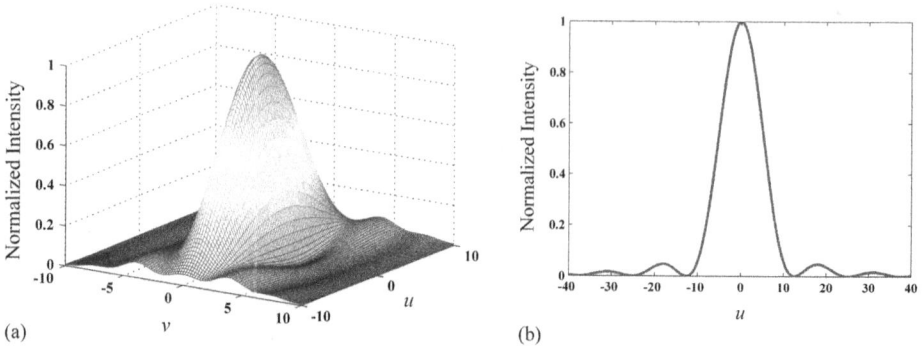

Figure 2.3. Three-dimensional and axial diffraction intensity distributions of a thin lens system. (a) 3D diffraction intensity distribution. (b) Axial intensity distribution.

$$I(v, u) = \left| \int_0^1 \exp\left(\frac{iu\rho^2}{2}\right) J_0(v\rho)\rho d\rho \right|^2. \tag{2.18}$$

Figure 2.3(a) describes the system's three-dimensional (3D) diffraction intensity distribution and figure 2.3(b) reveals the intensity distribution along the optical axis.

Considering the defocusing amount u, the 3D PSF $h(v, u)$ of a thin lens is defined as

$$h(v, u) = \int_0^1 \exp\left(\frac{iu\rho^2}{2}\right) J_0(v\rho)\rho d\rho \tag{2.19}$$

and the 3D IPSF is expressed as

$$I(v, u) = |h(v, u)|^2. \tag{2.20}$$

In a general microscopic imaging system, the point object's axial response curve can be obtained by integrating equation (2.20) with respect to v. Integral calculations show that an infinitely large detector can receive all diffractive signals. Although a CCD has a finite size physically, its size is much larger than an Airy disk, so the detector response with limited size and infinite integration response approximately coincides. However for CM imaging, not only the detector is required to have point-like size but also the illumination must be highly restricted to a scanning point. Prior to confocal imaging, flying spot imaging provided successful results in stray light control. However, the most interesting advance of CM was not to suppress cross talk, but open an era of 3D imaging [4, 5].

2.2 PSF in confocal imaging

The imaging models of CM are related to samples and they can be divided into transmissive and reflective types. Figure 2.4(a) and (b) are schematic diagrams of CM imaging systems working with transmissive and reflective samples, respectively.

In principle, transmissive and reflective confocal systems are not entirely equivalent. In industrial microscopic measurement, the sample itself is reflective, and so its imaging principle can be represented by figure 2.4(b). For biological fluorescence imaging, since the transparent sample is labelled with fluorescent

(a)

(b)

Figure 2.4. Schematic diagrams of CM imaging with transmissive and reflective samples. (a) Transmissive confocal microscopy. (b) Reflective confocal microscopy.

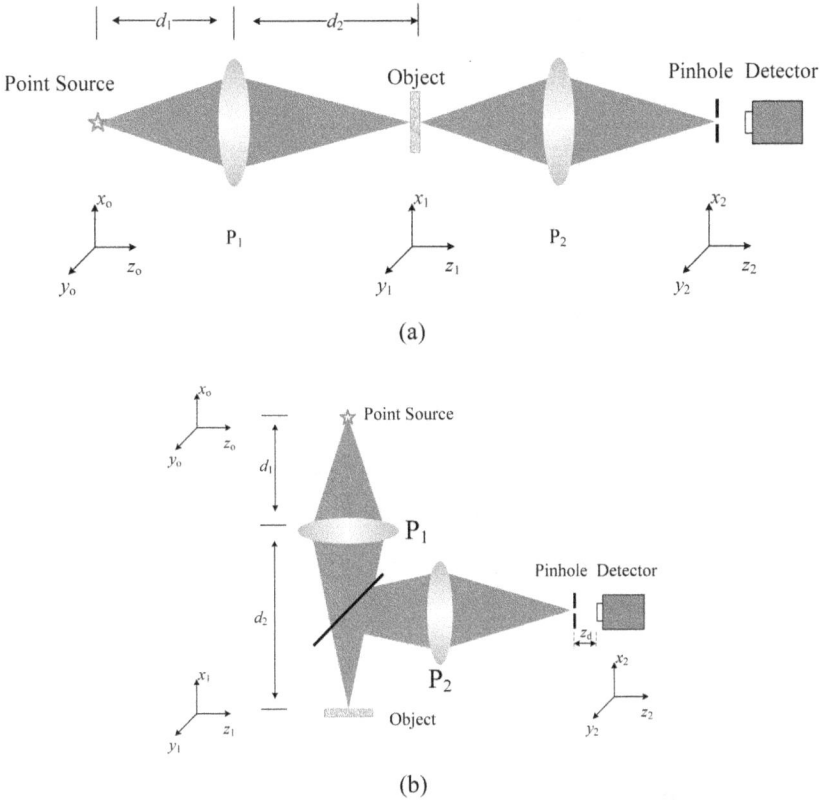

material, the system's measurement signal is the fluorescence signal rather than the transmitted optical field. In this case the imaging model can also be represented by the reflective system shown in figure 2.4(b). In this book we only discuss the reflective confocal imaging system.

As shown in figure 2.4, the light coming from a point source passes through the lens P_1 and converges on the measured surface, and the reflective light from the surface is then collected by the point detector with the help of a dichromic mirror and an additional lens P_2. In this system, the point source and the point detector are in conjugate positions. If $\delta(\mathbf{r}_o)$ denotes a point source, the amplitude distribution of the light field on the measured surface can be expressed as

$$U_{1\_\text{ill}}(r_1) = \int_{-\infty}^{\infty} \delta(r_0)h_1(r_0 + M_1 r_1)\mathrm{d}r_0 \qquad (2.21)$$

and

$$\mathbf{r}_0 = (x_0, y_0, z_0), \ \mathbf{r}_1 = (x_1, y_1, z_1), \ \mathbf{M}_1 = \begin{bmatrix} M_1 & 0 & 0 \\ 0 & M_1 & 0 \\ 0 & 0 & -M_1^2 \end{bmatrix} \text{ and } M_1 = d_1/d_2.$$

The reflected light field immediately behind the surface can be expressed as

$$U_{1\_ref}(\mathbf{r}_1) = U_{1\_ill}(\mathbf{r}_1)o(\mathbf{r}_s - \mathbf{r}_1) \tag{2.22}$$

where $\mathbf{r}_s = (x_s, y_s, z_s)$ and it represents the position of the scanning spot on the test plane. For the 3D amplitude PSF h_1 of lens P_2, the light field distribution on the imaging surface can be given by

$$
\begin{aligned}
U_2(\mathbf{r}_2, \mathbf{r}_s) &= \int_{-\infty}^{\infty} U_{1\_ref}(\mathbf{r}_1)h_1(\mathbf{r}_1 + \mathbf{M}_2\mathbf{r}_2)d\mathbf{r}_1 \\
&= \int_{-\infty}^{\infty}\left[\int_{-\infty}^{\infty}\delta(\mathbf{r}_0)h_1(\mathbf{r}_0 + \mathbf{M}_1\mathbf{r}_1)d\mathbf{r}_0\right] \cdot o(\mathbf{r}_s - \mathbf{r}_1)h_1(\mathbf{r}_1 + \mathbf{M}_2\mathbf{r}_2)d\mathbf{r}_1
\end{aligned} \tag{2.23}
$$

where

$$\mathbf{M}_2 = \begin{bmatrix} M_2 & 0 & 0 \\ 0 & M_2 & 0 \\ 0 & 0 & -M_2^2 \end{bmatrix} \text{ and } \mathbf{M}_2 = \mathbf{d}_2/\mathbf{d}_1$$

The position vector of the detector and the offset vector can be written as $\mathbf{r}_2 = (x_2, y_2, z_2)$ and $\mathbf{r}_{20} = (x_{20}, y_{20}, z_{20})$, respectively. The light intensity collected by the detector can be expressed as

$$
\begin{aligned}
I(\mathbf{r}_s) &= \int_{-\infty}^{\infty} |U_2(\mathbf{r}_2, \mathbf{r}_s)|^2 \, \delta(\mathbf{r}_2 - \mathbf{r}_{20})d\mathbf{r}_2 \\
&= \int_{-\infty}^{\infty}\left|\int\left[\int\delta(\mathbf{r}_0)\exp[ik(z_0 - z_1)]h_1(\mathbf{r}_0 + \mathbf{M}_1\mathbf{r}_1)d\mathbf{r}_0\right] \cdot o(\mathbf{r}_s - \mathbf{r}_1)\right. \\
&\quad \left. \times \exp[ik(z_1 - z_2)]h_1(\mathbf{r}_1 + \mathbf{M}_2\mathbf{r}_2)d\mathbf{r}_1\right|^2 \cdot \delta(\mathbf{r}_2 - \mathbf{r}_{20})d\mathbf{r}_2
\end{aligned} \tag{2.24}
$$

Equation (2.24) can be further simplified to

$$I(\mathbf{r}_s) = |h_a(\mathbf{r}_s) \otimes_3 o(\mathbf{r}_s)|^2 \tag{2.25}$$

and

$$h_a(\mathbf{r}) = \exp[ik(-z \pm z)]h_1(\mathbf{M}_1\mathbf{r})h_2(\mathbf{r} + \mathbf{M}_2\mathbf{r}). \tag{2.26}$$

$\mathbf{r}(x, y, z)$ is the position vector on the object plane. In a rotational symmetric system, if a lens's radius is a, the PSF can be expressed as

$$h_a(v_x, v_y, u) = h_1(v, u)h_2(v_x, v_y, u) \tag{2.27}$$

and $v = \sqrt{v_x^2 + v_y^2}$. When the sample is in focus and the detector is out of focus, the PSFs of the illuminating arm and the detecting arm can be given by

$$h_1(v, u) = \exp(-is_0u)\int_0^1 P_1(\rho, u)J_0(v\rho)\rho d\rho \tag{2.28}$$

$$h_2(v, u) = \exp(-is_0u)\int_0^1 P_2(\rho, u)\exp(iu\rho^2/2)J_0(v\rho)\rho d\rho \tag{2.29}$$

where the non-dimensional coordinates of an object space can be expressed as

$$v_{(v_x, v_y)} = \frac{2\pi}{\lambda} \frac{a}{d_2} r_{(x,y)}, \quad u = \frac{2\pi}{\lambda} \frac{a^2}{d_2^2} z \tag{2.30}$$

and the non-dimensional coordinates of the image space can be expressed as

$$(v_2, v_{2x}, v_{2y}) = \frac{2\pi}{\lambda} \frac{a}{d_1} (r_2, x_2, y_2), \quad u_2 = \frac{2\pi}{\lambda} \frac{a^2}{d_1^2} z_2. \tag{2.31}$$

When the measured object is a point object, and the detector is in focus ($u_2 = 0$), from equations (2.25), (2.27), (2.28), and (2.29), we can obtain the 3D-IPSF expressed as

$$I(\mathbf{r}_s) = |h_a(\mathbf{r}_s) \otimes_3 \delta(\mathbf{r}_s)|^2 = |h_1(v, u) \cdot h_2(v, u)|^2 = |h(v, u)|^4 \tag{2.32}$$

and

$$h(v, u) = C \int_0^1 P_1(\rho, u) J_0(v\rho) \rho \, d\rho. \tag{2.33}$$

If $u = 0$, the normalized expression of the lateral light intensity distribution can be expressed as

$$I(v) = \left(\frac{2J_1(v)}{v} \right)^4. \tag{2.34}$$

For comparison, the corresponding intensity distribution in a traditional optical microscope can be expressed as $I(v) = [2J_1(v)/v]^2$.

If $v = 0$, the normalized expression of the axial light intensity distribution can be shown as follows:

$$I(u) = \left[\frac{\sin(u/4)}{u/4} \right]^4. \tag{2.35}$$

By comparing the diffraction spot transverse curves in figures 2.3 and 2.5, we find that the full width at half maximum (FWHM) of CM's intensity response in optical coordinates is 1.32 and the width of traditional microscopy's is 3.23. So the lateral

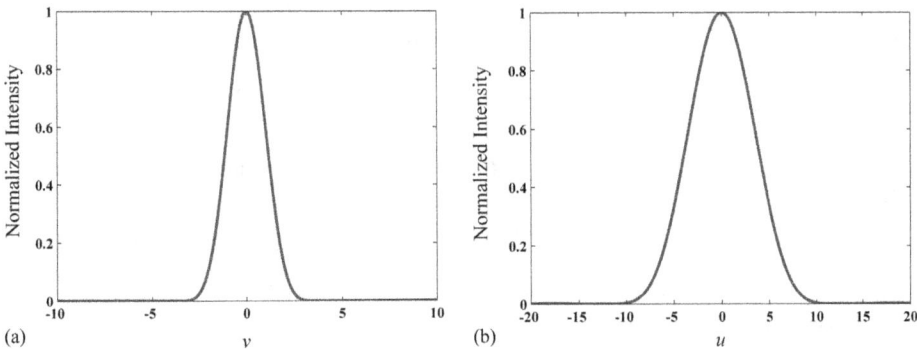

Figure 2.5. Lateral and axial characteristic curves of a confocal microscopy system. (a) Lateral response. (b) Axial response.

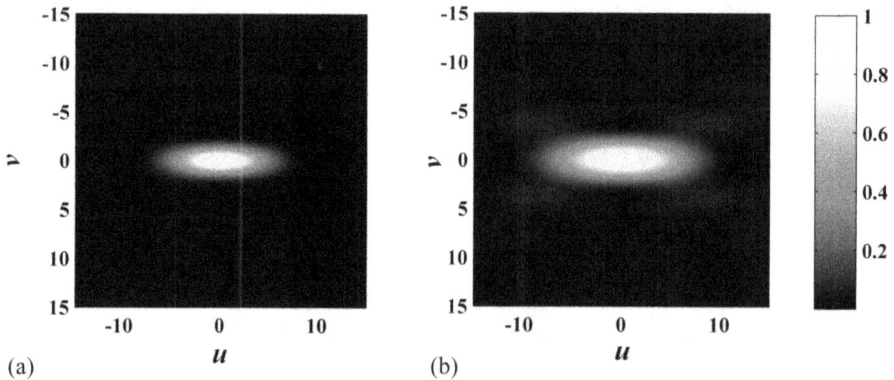

Figure 2.6. 3D light intensity distributions of (a) CM and (b) traditional microscopy.

resolution of CM is 1.4 times that of traditional microscopy. It can be seen from figures 2.4 and 2.5 that the axial response curve of CM is sharper than that of traditional microscopy.

It can be seen from figure 2.6(a) that the pinhole-caused suppression of the IPSF severely changes the sectioning property of CM, resulting in its wide usage in microscopic metrology and biological imaging in 3D applications [6, 8, 9].

PSF is a direct way to evaluate the resolution abilities of different imaging systems. However, it is easy to overlook the impact of the side lobe spot. Another more accurate way to evaluate the imaging capability of an optical system is to calculate and analyze the system's transfer function, so we can understand better how the main and side lobe of the PSF of an imaging system affects its imaging contrast. In a subsequent chapter of this book, the transfer function of CM will be further discussed.

References

[1] Born M E 1980 *Wolf Principles of Optics* vol 6 (New York: Pergamon)

[2] Goodman J W 2005 *Introduction to Fourier optics* (New York: Roberts and Company Publishers)

[3] Hecht E 1987 *Optik* (New York: McGraw-Hill)

[4] Gu M and Gan X S 1996 Fresnel diffraction by circular and serrated apertures illuminated with an ultrashort pulsed-laser beam *J. Opt. Soc. Am.* A **13** 771–8

[5] Gu M 1995 Three-dimensional space-invariant point-spread function for a single lens *J. Opt. Soc. Am.* A **12** 1602–4

[6] Wilson T and Sheppard C 1984 *Theory and Practice of Scanning Optical Microscopy* (London: Academic Press)

[7] Gu M 2000 *Advanced optical imaging theory* (New York: Springer Science and Business Media)

[8] Gu M 1996 *Principles of Three Dimensional Imaging in Confocal Microscopes* (Singapore: World Scientific)

[9] Sheppard C J R and Shotton D M 1997 *Confocal Laser Scanning Microscopy* (Oxford: BIOS Scientific Publishers)

IOP Concise Physics

Confocal Microscopy

Mengzhou Li, Hongting Wang, Jian Liu and Jiubin Tan

Chapter 3

Incoherent three-dimensional optical transfer function

3.1 Development of 3D optical transfer function

Optical transfer function (OTF) theory developed from Fourier optics, and was originally proposed in the 1950s' literature of H H Hopkins. By discussion and agreement during several international optical conferences, the term 'optical transfer function' was finally adopted. With the rapid development of computer techniques, OTF theory has been widely accepted and applied.

In 1966, B R Frieden was the first to put forward the concept of the three-dimensional optical transfer function (3D-OTF). Frieden introduced the 'axial' dimension into the conventional 'lateral' transfer function and wrote the 3D description of the spatial frequency spectrum of weak scattering sample imaging [1]. 3D-OTF includes 3D descriptions of samples and theoretically can be used to depict the imaging process of any sample. Only if the optical system satisfies space invariance conditions, is 3D-OTF able to overcome the limitation of the conventional transfer function, which is used to describe the space frequency distribution on the focal plane. Moreover, the 2D transfer function can easily be obtained through the axial Fourier transform of the 3D transfer function.

After Frieden's discussions regarding the 3D coherent transfer function [1], E Wolf established OTF theory under incoherent illumination [2], and N Streibl carried out a considerable amount of research into the OTF of partially coherent imaging [3].

Since the 1980s, with the development of 3D microscopic imaging techniques, OTF theory has emerged as an important tool in studying the sectional abilities of optical systems. In the 1990s, T Wilson, C J R Sheppard and M Gu published a series of articles comprising in-depth discussions of the 3D imaging properties of confocal microscopy (CM) [4–6]. They investigated the influence of pupil modulation and detector size on confocal imaging properties [7–9], and studied the straight edge

doi:10.1088/978-1-6817-4337-0ch3 © Morgan & Claypool Publishers 2016

response of CM [10, 11]. In 2000, *Three-Dimensional Imaging in Confocal Microscopes*, by M Gu, was translated into Chinese and was widely read in China.

3.2 3D imaging models of CM

The imaging process is coherent when a coherent light source is used and the image formation is the superposition of amplitudes. Industrial confocal scanning microscopes which commonly use a laser as the light source, belong to the coherent imaging type. Usually, by placing a pinhole in front of an expanding incoherent light source, we can obtain a coherent light source. When fluorescence is used, the imaging process is incoherent and demonstrates intensity combination. Biological confocal scanning microscopes, which commonly use fluorescent dyes for imaging, belong to this incoherent imaging type. When an incoherent light source like a halogen lamp and a finite-sized pinhole are used, the imaging process is partially coherent.

Figure 3.1 is a schematic diagram of a reflection mode scanning CM. An extended incoherent light and an intensity detector are used. $S(x, y)$ and $D(x, y)$ are distribution functions of illumination intensity and detection sensitivity, respectively.

For partially coherent imaging formation, the method is to firstly calculate the intensity response on the image plane, of every point of the light source plane, and then calculate the image response, by combining the intensity distributions of those points due to the incoherence of said points.

When incoherent illumination and a finite-sized pinhole are used, the detected optical signal can be depicted as

$$I_{\text{img}}(x_s, y_s, z_s) = \iint_{-\infty}^{\infty} D(x_5, y_5) \iint_{-\infty}^{\infty} S(x_1, y_1) \left| \iiint_{-\infty}^{\infty} \exp(-2jkz_3) \right.$$

$$\times h_1\left(x_1 - \frac{x_3}{M}, y_1 - \frac{y_3}{M}, -\frac{z_3}{M^2}\right) h_1\left(x_5 - \frac{x_3}{M}, y_5 - \frac{y_3}{M}, -\frac{z_3}{M^2}\right) \quad (3.1)$$

$$\left. \times t(x_3 - x_s, y_3 - y_s, z_3 - z_s) \mathrm{d}x_3 \mathrm{d}y_3 \mathrm{d}z_3 \right|^2 \mathrm{d}x_1 \mathrm{d}y_1 \mathrm{d}x_5 \mathrm{d}y_5$$

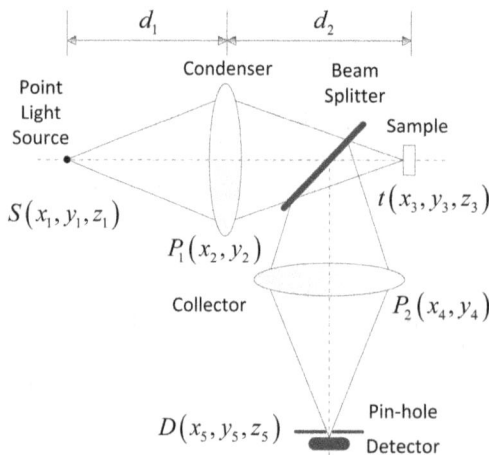

Figure 3.1. Schematic diagram of reflection mode scanning CM.

where h_1 and h_2 represent the imaging process of illumination and collecting respectively, and usually h_1 equals h_2, and $h_1(x, y, z)$ is defined as the 3D amplitude point spread function (PSF) of a single lens as

$$h_1(x, y, z) = \iint_{-\infty}^{\infty} P(x_2, y_2) \exp\left[-\frac{jk}{2d_{10}^2} z(x_2^2 + y_2^2) \right]$$

$$\times \exp\left[\frac{jk}{d_{10}} (xx_2 + yy_2) \right] dx_2 dy_2. \tag{3.2}$$

When a point source and a point detector are used, equation (3.1) can be rewritten as

$$I_{\text{img}}(x_s, y_s, z_s) = \left| \iint_{-\infty}^{\infty} \exp(-2jkz_3) h_1\left(-\frac{x_3}{M}, -\frac{y_3}{M}, -\frac{z_3}{M^2} \right) \right.$$

$$t(x_3 - x_s, y_3 - y_s, z_3 - z_s)$$

$$\left. \times h_2\left(-\frac{x_3}{M}, -\frac{y_3}{M}, -\frac{z_3}{M^2} \right) dx_3 dy_3 dz_3 \right|^2$$

$$= \left| \left[t(-x_s, -y_s, -z_s) \exp(2jkz_s) \right] \right.$$

$$\left. \otimes \left[h_1\left(-\frac{x_3}{M}, -\frac{y_3}{M}, -\frac{z_3}{M^2} \right) h_2\left(-\frac{x_3}{M}, -\frac{y_3}{M}, -\frac{z_3}{M^2} \right) \right] \right|^2. \tag{3.3}$$

It can be seen from equation (3.3) that with an ideal pinhole, a partially coherent situation can be transformed into a completely coherent situation. Therefore, in real cases, if the size of the pinhole used is small enough, the coherent model can be used directly for analysis regardless of the type of light source chosen.

For the process involving fluorescence, we ignored the wavelength difference between the illumination light and the excited fluorescence. The basic calculation method is to divide them into two parts, each part being an incoherent imaging process. Firstly, the 3D light intensity distribution on the object's space is

$$I_3(x_3, y_3, z_3) = \iint_{-\infty}^{\infty} S(x_1, y_1) \left| h_1\left(x_1 - \frac{x_3}{M}, y_1 - \frac{y_3}{M}, -\frac{z_3}{M^2} \right) \right|^2 dx_1 dy_1. \tag{3.4}$$

Then the intensity signal detected by the detector is

$$I_{\text{img}}(x_s, y_s, z_s) = \iint_{-\infty}^{\infty} D(x_5, y_5) \iiint_{-\infty}^{\infty} I_3(x_3, y_3, z_3) t(x_3 - x_s, y_3 - y_s, z_3 - z_s)$$

$$\times \left| h_1\left(x_5 - \frac{x_3}{M}, y_5 - \frac{y_3}{M}, -\frac{z_3}{M^2} \right) \right|^2 dx_3 dy_3 dz_3 dx_5 dy_5. \tag{3.5}$$

When a point source and a point detector are used, equation (3.5) can be rewritten as

$$I_{img}(x_s, y_s, z_s) = \iiint_{-\infty}^{\infty} \left| h_1\left(-\frac{x_3}{M}, -\frac{y_3}{M}, -\frac{z_3}{M^2}\right) \right|^2 t(x_3 - x_s, y_3 - y_s, z_3 - z_s)$$

$$\times \left| h_2\left(-\frac{x_3}{M}, -\frac{y_3}{M}, -\frac{z_3}{M^2}\right) \right|^2 dx_3 dy_3 dz_3 \tag{3.6}$$

$$= t(-x_s, -y_s, -z_s)$$

$$\otimes \left| h_1\left(-\frac{x_s}{M}, -\frac{y_s}{M}, -\frac{z_s}{M^2}\right) h_2\left(-\frac{x_s}{M}, -\frac{y_s}{M}, -\frac{z_s}{M^2}\right) \right|^2.$$

Furthermore, when the object is an ideal point, the 3D intensity PSF of confocal incoherent imaging can be obtained and given by

$$I_h(x_s, y_s, z_s) = \left| h_1\left(-\frac{x_s}{M}, -\frac{y_s}{M}, -\frac{z_s}{M^2}\right) h_2\left(-\frac{x_s}{M}, -\frac{y_s}{M}, -\frac{z_s}{M^2}\right) \right|^2. \tag{3.7}$$

3.3 3D-OTF of CM

In section 3.2, 3D image formation in CM was introduced under the incoherent situation. With a point source and a point detector, the 3D intensity PSF of CM can be depicted using equation (3.7). Equation (3.2) can be rewritten as equation (3.8), when the actual coordinates are transformed into optical coordinates using $v = \sin\alpha \cdot x/\lambda$, $w = \sin\alpha \cdot y/\lambda$, and $u = 4kz \cdot \sin^2(\alpha/2)$, where α is the half aperture angle of the objective,

$$h_1(v, w, u) = \iint_{-\infty}^{\infty} P(x_2, y_2) \exp\left[-j\frac{u}{2}\left(\frac{x_2^2}{a^2} + \frac{y_2^2}{a^2}\right)\right]$$

$$\times \exp\left[j2\pi\left(v\frac{x_2}{a} + w\frac{y_2}{a}\right)\right] dx_2 dy_2. \tag{3.8}$$

By normalizing the horizontal and vertical coordinates with $\xi = x_2/a$ and $\eta = y_2/a$, and omitting the constant-coefficient, we get

$$h_1(v, w, u) = \iint_{-\infty}^{\infty} P(\xi, \eta) \exp\left[-j\frac{u}{2}(\xi^2 + \eta^2)\right] \exp[j2\pi(v\xi + w\eta)] d\xi d\eta \tag{3.9}$$

where

$$P(\xi, \eta) = \begin{cases} 1 & \xi, \ \eta \in (0, 1) \\ 0 & \text{others} \end{cases}.$$

By ignoring the amplification and reversal of coordinates, equation (3.7) can be written as

$$I_h(v, w, u) = |h_1(v, w, u)|^2 |h_2(v, w, u)|^2. \tag{3.10}$$

Through the Fourier transform of equation (3.10), the 3D-OTF of CM can be given by

$$C(l_m, l_n, s) = F_3\{| h_1 |^2 \cdot | h_2 |^2\} = F_3\{| h_1 |^2\} \otimes F_3\{| h_2 |^2\}, \tag{3.11}$$

where l_m, l_n, s are the spatial frequencies of v, w, u respectively, and F_3 stands for the 3D Fourier transform operation.

The 3D optical pupil function can be defined as

$$P(l, u) = \begin{cases} \exp\left(-i\dfrac{u}{2}l^2\right) & l \in (0, 1) \\ 0 & \text{others} \end{cases} \tag{3.12}$$

where $l = \sqrt{l_m^2 + l_n^2}$, $C_{CM}(l, s) = C(l_m, l_n, s)$, $h = h_1 = h_2$, $C'(l, s) = F_3\{| h |^2\}$.

In accordance with the properties of the Fourier transform and equation (3.9), we have $F_{v, w}\{h(v, w, u)\} = P(l, u)$, thus

$$C'(l, s) = \int_{-\infty}^{\infty} \left[P(l, u) \otimes P^*(l, u) \right] \exp(-i2\pi us) du. \tag{3.13}$$

Supposing $g(l_m, l_n, u) = P(l, u) \otimes P^*(l, u)$, it can be written as

$$\begin{aligned} g(l_m, l_n, u) = &\iint_A \exp\left[i\frac{u}{2}(x^2 + y^2) \right] \\ &\times \exp\left\{ -i\frac{u}{2}\left[(l_m - x)^2 + (y - l_n)^2 \right] \right\} dxdy. \end{aligned} \tag{3.14}$$

The geometric meaning of equation (3.14) is the integral of the overlapping area of the two circles, with the weighting factor

$$\exp\left[i\frac{u}{2}(x^2 + y^2) \right] \exp\left\{ -i\frac{u}{2}\left[(l_m - x)^2 + (y - l_n)^2 \right] \right\}. \tag{3.15}$$

Due to the fact that the value of the weighting factor will not change while the relative position of the two circles remain unchanged, a coordinate system can be re-established by taking the line joining the two centers as the horizontal axis, and the midpoint of two centers as the origin, as shown in figure 3.2. Thus, $g(l, u)$ can be calculated through the situation $l_m = l, l_n = 0$, and the equation (3.14) can be written as

$$\begin{aligned} g(l, u) &= \iint_{A'} \exp\left[i\frac{u}{2}\left[(x + 0.5l)^2 + y^2 \right] \right] \exp\left\{ -i\frac{u}{2}\left[(x - 0.5l)^2 + y^2 \right] \right\} dxdy \\ &= \iint_{A'} \exp(iuxl) dxdy. \end{aligned} \tag{3.16}$$

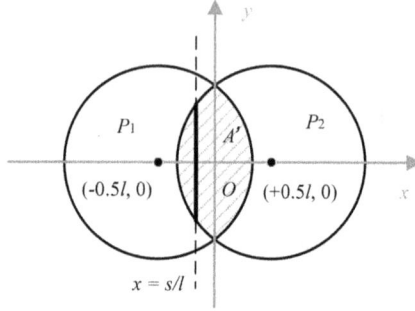

Figure 3.2. An illustration for two-dimensional convolution.

Using equations (3.16) and (3.13), we can obtain

$$C'(l, s) = \int_{-\infty}^{\infty} \iint_{A'} \exp(iuxl) \exp(-i2\pi us) dx dy du$$
$$= \iint_{A'} \delta(s - lx) dx dy. \tag{3.17}$$

The geometric meaning of equation (3.17) is the segment length of the part of line $x = s/l$ (the segment line in bold), which is in the overlapping area A'. We can obtain

$$C'(l, s) = \frac{2}{l^2} \operatorname{Re}\left\{ \sqrt{l^2 - \left(\frac{l^2}{2} + |s|\right)^2} \right\}. \tag{3.18}$$

By substituting formula (3-8) into formula (3-11) and calculating the convolution with formula (3-16) we can obtain the 3D-OTF of CM,

$$C_{CM}(l, s) = C'(l, s) \otimes C'(l, s)$$

$$= \iiint_{-\infty}^{\infty} \frac{4}{l_1^2 l_2^2} \left\{ \operatorname{Re}\left[\sqrt{l_1^2 - \left(\frac{l_1^2}{2} + |s' - 0.5s|\right)^2} \right] \right\}$$

$$\times \left\{ \operatorname{Re}\left[\sqrt{l_2^2 - \left(\frac{l_2^2}{2} + |s' + 0.5s|\right)^2} \right] \right\} dx dy ds' \tag{3.19}$$

where $l_1^2 = (x + 0.5l)^2 + y^2$, $l_2^2 = (x - 0.5l)^2 + y^2$.

3.4 3D-OTF of differential CM

Confocal pinholes are introduced in CM to filter out the out of focus light signals. Therefore, this technique not only inherits the non-contact advantage of optical measurement over the method using a 3D coordinate measuring machine, but also has the features of high anti-stray-light ability, high intensity contrast, and axial sectioning ability. Thus, the technique is widely used for 3D measurement.

Differential confocal microscopy (DCM) further overcomes some shortcomings of conventional CM, including the insensitivity to the surface profile height in the peak position, the poor ability to resist the influences of optical intensity fluctuation, the electronic noise of detectors, and the inability to achieve bipolar absolute tracking on discontinuous surface profiles [12, 13]. Furthermore the axial sensitivity of a confocal system is significantly improved using differential techniques.

The main difference between DCM and conventional CM lies in the detection part. In conventional CM, the detector is placed on the focus plane of the collecting lens, while two symmetrically defocused detectors are used in DCM. The differential signal is obtained by the subtraction of the signals from the two detectors. The basic optical layout of DCM is shown in figure 3.3.

As shown in figure 3.3, the differential optical path can be regarded as two independent paths. Each of them is generally consistent with conventional CM, only if the detectors in DCM are defocused at position $\pm z_d$. For the incoherent situation, the 3D intensity PSF of each path can be depicted as

$$I_h(x_s, y_s, z_s) = \left| h_1\left(-\frac{x_s}{M}, -\frac{y_s}{M}, -\frac{z_s}{M}\right) \right|^2 \left| h_1\left(-\frac{x_s}{M}, -\frac{y_s}{M}, -\frac{z_s \pm M^2 z_d}{M^2}\right) \right|^2. \quad (3.20)$$

The 3D intensity PSF of DCM can be obtained using

$$I_{hd}(x_s, y_s, z_s) = \left| h_1\left(-\frac{x_s}{M}, -\frac{y_s}{M}, -\frac{z_s}{M^2}\right) \right|^2 \left| h_1\left(-\frac{x_s}{M}, -\frac{y_s}{M}, -\frac{z_s - M^2 z_d}{M^2}\right) \right|^2$$
$$- \left| h_1\left(-\frac{x_s}{M}, -\frac{y_s}{M}, -\frac{z_s}{M^2}\right) \right|^2 \left| h_1\left(-\frac{x_s}{M}, -\frac{y_s}{M}, -\frac{z_s + M^2 z_d}{M^2}\right) \right|^2. \quad (3.21)$$

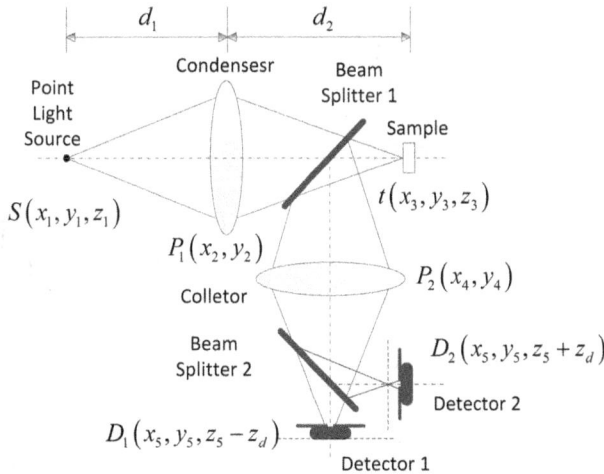

Figure 3.3. General optical layout of a reflection mode DCM.

Through normalization of coordinates, equation (3.21) becomes

$$I_{h\_d}(v, w, u) = |h(v, w, u)|^2 |h(v, w, u - u_d)|^2$$
$$- |h(v, w, u)|^2 |h(v, w, u + u_d)|^2. \tag{3.22}$$

Through the 3D Fourier transform, 3D-PSF can be transformed into 3D-OTF as

$$C_{DCM}(l, s) = C_{DCM}(l_m, l_n, s) = F_3[I_{h\_d}(v, w, u)], \tag{3.23}$$

where $l = \sqrt{l_m^2 + l_n^2}$.

Based on the properties of the Fourier transform operation, it can be found that $F_{v,w}[h] = P(l, u)$, $F_3[|h|^2] = F_s\{P(l, u) \otimes P^*(l, u)\}$, where $l = \sqrt{l_m^2 + l_n^2}$. When h is $h(v, w, u)$, we have $P(l, u) \otimes P^*(l, u) = \iint_{A'} \exp(iulx)\mathrm{d}x\mathrm{d}y$ and

$$F_3\left[|h(v, w, u)|^2\right] = \underset{u \to s}{F} \left\{ \iint_{A'} \exp(iulx)\mathrm{d}x\mathrm{d}y \right\}$$
$$= \iint_{A'} \underset{u \to s}{F} [\exp(iulx)]\mathrm{d}x\mathrm{d}y \tag{3.24}$$
$$= \frac{1}{l} \int_{x=\frac{s}{l}} \mathrm{d}y = \frac{2}{l^2} \mathrm{Re}\left[\sqrt{l^2 - \left(|s| + \frac{l^2}{2}\right)^2} \right].$$

Similarly, when h is $h(v, w, u + u_d)$, $u + u_d$ can be seen as a whole as u_d is a constant. Thus, we can obtain

$$F_3\left[|h(v, w, u + u_d)|^2\right] = \underset{u \to s}{F} \left\{ \iint_{A'} \exp(iulx)\exp(iu_dlx)\mathrm{d}x\mathrm{d}y \right\}$$
$$= \iint_{A'} \underset{u \to s}{F}\left[\exp(iulx)\right]\exp(iu_dlx)\mathrm{d}x\mathrm{d}y$$
$$= \iint_{A'} \exp(iu_dlx)\delta(s - lx)\mathrm{d}x\mathrm{d}y \tag{3.25}$$
$$= \exp(iu_ds)\frac{1}{l} \int_{x=\frac{s}{l}} \mathrm{d}y$$
$$= \exp(iu_ds)\frac{2}{l^2} \mathrm{Re}\left[\sqrt{l^2 - \left(|s| + \frac{l^2}{2}\right)^2} \right].$$

When h is $h(v, w, u-u_d)$, we can obtain

$$F_3\left[|h(v, w, u - u_d)|^2\right] = \exp(-iu_ds)\frac{2}{l^2} \mathrm{Re}\left[\sqrt{l^2 - \left(|s| + \frac{l^2}{2}\right)^2} \right]. \tag{3.26}$$

For better illustration, supposing $C_1 = F_3[|h(v, w, u)|^2] \otimes F_3[|h(v, w, u - u_d)|^2]$ and $C_2 = F_3[|h(v, w, u)|^2] \otimes F_3[|h(v, w, u + u_d)|^2]$, then we have $C_{dcm} = C_1 - C_2$, that is

$$C_{dcm} = F_3[|h(v, w, u)|^2]$$
$$\otimes \left\{ F_3[|h(v, w, u - u_d)|^2] - [|h(v, w, u + u_d)|^2] \right\}. \tag{3.27}$$

From equations (3.25) and (3.26), we have

$$F_3\left[|h(v, w, u - u_d)|^2\right] - F_3\left[|h(v, w, u + u_d)|^2\right]$$
$$= \frac{-4i \sin(u_d s)}{l^2} \mathrm{Re}\left[\sqrt{l^2 - \left(|s| + \frac{l^2}{2}\right)^2} \right]. \tag{3.28}$$

Finally we can obtain the expression of the OTF of DCM

$$C_{DCM} = \left\{ \frac{2}{l^2} \mathrm{Re}\left[\sqrt{l^2 - \left(|s| + \frac{l^2}{2}\right)^2} \right] \right\}$$
$$\otimes \left\{ \frac{-4i \sin(u_d s)}{l^2} \mathrm{Re}\left[\sqrt{l^2 - \left(|s| + \frac{l^2}{2}\right)^2} \right] \right\}. \tag{3.29}$$

The simulation of the PSF and the transfer function of the CM and DCM with a different defocusing amount is shown in figure 3.4. Usually in CM, the peak point is chosen for sample surface reconstruction. While in DCM, the zero point is used to reconstruct the sample surface, or the linear zone near zero is used to map its height [14]. The DCM PSF distribution varies with the defocusing amount; when the amount is sufficiently large, the behavior is no longer linear, as shown in figure 3.4(c). Linear-zone gradients are compared to determine the maximum sensitivity. CM PSFs do not have negative values, whereas DCM PSFs have both positive and negative values. The CM OTF is symmetric about the origin, and the cut-off frequencies for the axial and lateral directions are 1 and 4, respectively. The DCM OTF magnitudes are symmetric about the two coordinate axes, and their cut-off frequencies are identical to those of CM OTFs. Fluctuations appear when the defocusing amount is large, and result in image distortion.

This chapter gives a brief introduction to the basic theory of the 3D transfer function and introduces the application of the 3D transfer function in compound detection property analysis by using DCM as an example. It should be pointed out that transfer function theory can only be applied in the transfer property analysis of a linear optical system rather than a nonlinear optical system. As for the nonlinear super-resolution imaging technique, it would be hard to discuss its transfer property through bandwidth analysis. System information bandwidth, can be used to evaluate the system information transmission capacity, thus it is a more effective method of

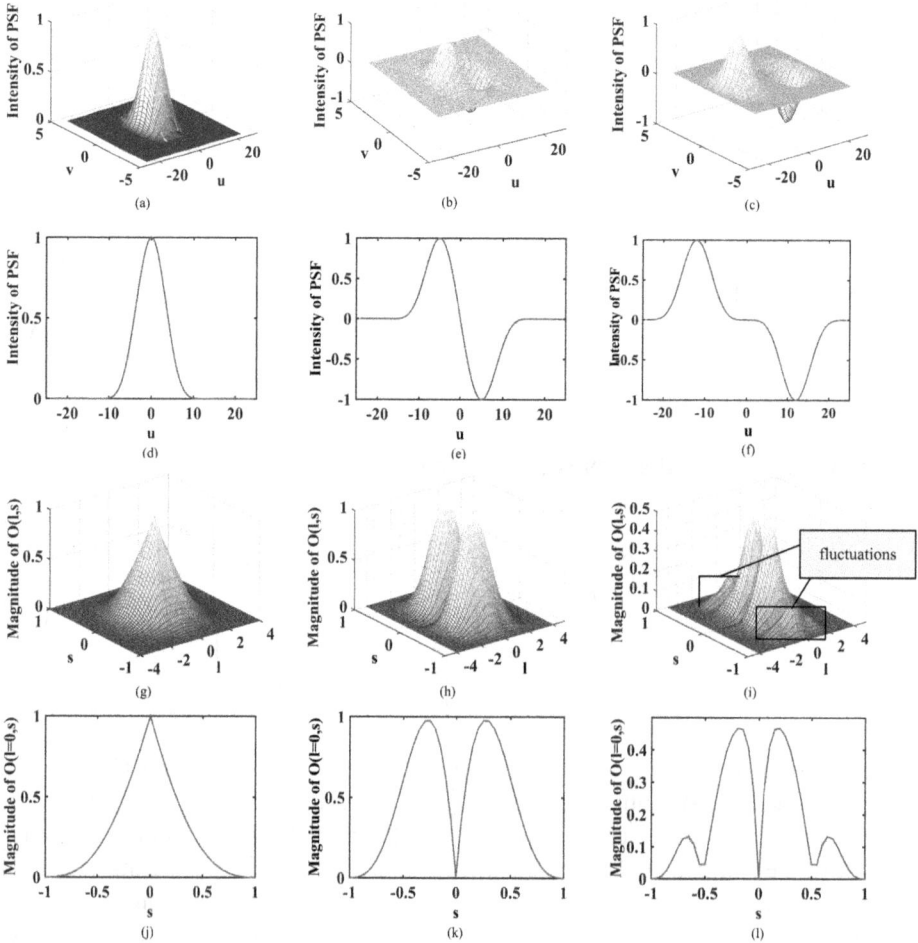

Figure 3.4. PSFs and magnitudes of OTFs for CM and DCM with various defocusing amounts: (a) 3D PSF for CM; (b) 3D PSF for DCM with $u_d = 5$; (c) 3D PSF for DCM with $u_d = 12$; (d) axial PSF for CM; (e) axial PSF for DCM with $u_d = 5$; (f) axial PSF for DCM with $u_d = 12$; (g) magnitude of the 3D OTF for CM; (h) magnitude of the 3D OTF with $u_d = 5$; (i) magnitude of the 3D OTF with $u_d = 12$; (j) magnitude of the axial OTF for CM; (k) magnitude of the axial OTF with $u_d = 5$; and (l) magnitude of the axial OTF with $u_d = 12$. Reproduced with permission from Tan J, Wang H, Li M, and Liu J 2015 *J. Microsc.* **261** 259. Copyright 2015 John Wiley and Sons.

reflecting imaging performance. System property analysis of nonlinear optical imaging may become one of the foci of future basic optical research.

References

[1] Freieden B R 1967 Optical transfer of the 3D object *J. Opt. Soc. Am.* **57** 56–65
[2] Born M and Wolf E 2000 Principles of optics: electromagnetic theory of propagation, interference and diffraction of light *CUP Archive*

[3] Streibl N 1985 3D imaging by a microscope *J. Opt. Soc. Am.* A **2** 121–7

[4] Sheppard C J R, Gu M and Kawata Y 1994 3D transfer functions for high-aperture systems *J. Opt. Soc. Am.* A **11** 593–8

[5] Sheppard C J R and Gu M 1992 The significance of 3_D transfer functions in confocal scanning microscopy *J. Microsc.* **165** 377–90

[6] Gu M and Sheppard C J R 1992 Effects of defocus and primary spherical aberration on 3D coherent transfer functions in confocal microscopes *Appl. Opt.* **31** 2541–9

[7] Wilson T and Carlini A R 1987 Size of the detector in confocal imaging systems *Opt. Lett.* **12** 227–9

[8] Gu M and Sheppard C J R 1992 3D coherent transfer function in reflection-mode confocal microscopy using annular lenses *J. Mod. Opt.* **39** 783–793

[9] Gu M 1996 *Principles of Three Dimensional Imaging in Confocal Microscopes* (Singapore: World Scientific)

[10] Sheppard C J R and Heaton J M 1984 Confocal images of straight edges and surface steps *Optik* **68** 371–80

[11] Sheppard C J R and Hamilton D K 1984 Edge enhancement by defocusing of confocal images *J. Mod. Opt.* **31** 723–7

[12] Liu J, Tan J, Bin H and Wang Y 2009 Improved differential confocal microscopy with ultrahigh signal-to-noise ratio and reflectance disturbance resistibility *Appl. Opt.* **3** 476–84

[13] Tan J, Liu J and Wang Y 2010 Differential confocal microscopy with a wide measuring range based on polychromatic illumination *Meas. Sci. Technol.* **21** 054013

[14] Tan J, Wang H, Li M and Jian L 2015 3D optical transfer function in differential confocal microscopy *J. Microsc.* **261** 259–66

Chapter 4

Decoupling criteria for three-dimensional optical microscopic measurement

4.1 Introduction

In the field of metrology, optical stereo microscopic imaging and measuring instruments have been widely applied for the measurement of optical component surface profiles [1], integrated circuit lines [2], width and geometrical parameters of microelectromechanical devices [3], and alignment of microstructural encapsulation [4]. The difference between three-dimensional (3D) microscopic measurement and traditional 2D microscopic imaging lies in the fact that the latter focuses on lateral resolution and contrast, while the former not only requires a good capability of lateral resolution but also an accurate height of sample. With the introduction of height, the technologies and theories of microscopic instruments become more complicated. Meanwhile a coupling effect appears in 3D measurement, which has drawn much worldwide attention from the research community, including the International Standardization Organization. The coupling effect in 3D measurement made with optical microscopy is subject to a phenomenon regarding the accuracy of height measurement of a groove or step sample, which suffers a principle error caused by the influence of the relatively small transverse period of the sample. Users of the instrument may find that the height measurement is influenced by the lateral period of samples. Under the same height, the smaller the lateral period is, the more serious the height measurement deviation is, which cannot be explained using current theories of optical resolution.

In 2007, PTB in Germany reported a method which can be used to evaluate 3D response characteristics of the scanning probe microscope with a pyramid step structure [5]. The user can evaluate the deviation of height measurement through the measurement of the sample. However how to avoid the deviation has not been explained in theories. In the process of international standardization of measurement methods, the demands of 3D microscopic measurement accuracy continually

doi:10.1088/978-1-6817-4337-0ch4

© Morgan & Claypool Publishers 2016

increase and the error margin in height measurement is required to be nanometers or even smaller. Thus, whether common users of instruments or professional researchers in metrology, all have an urgent demand for criteria to assess effective 3D resolution of microscope systems and prediction of measurement validity, in order that proper parameters of the instrument and portions of the sample for height assessment can be selected, to avoid the principle error in 3D measurement.

The W/3 method is widely adopted in current international and national standards and in metrology criteria. The W/3 method states that the upper surface on each side of a groove shall be ignored for a length equal to one-third of the width of the groove. The surface at the bottom of the groove is assessed only over the central third of its width [6]. This method of reading the height value can be used to avoid the influence of any rounding of the corners in stylus measurement. Consequently, it is applied widely in this application. However the W/3 principle is not suitable for direct application in optical instruments, because the resolution of an optical system is not only relevant to sample characteristics but also to systematic parameters such as illumination wavelength and numerical aperture. The W/3 method does not reflect the relevance between sample characteristics and the representational ability of optical instruments, nor the influence of step height or groove depth on measurement.

Decoupling models of optical microscopic 3D measurements of thin samples and deep groove samples have revealed the relevance between sample characteristics and the representational ability of optical instruments. These models are based on the convolution irrelevance principle (CIP) and limited energy loss condition through the interaction of the point spread function (PSF) of optical systems and object function. The relationships between the coupling distances of measured samples and the heights of steps, illumination wavelength, and the numerical apertures of objectives are depicted in these models to provide a new theoretical reference for 3D structural representation of groove or step samples.

4.2 Decoupling model for the measurement of thin samples

The common methods of optical 3D measurement can be divided into wide field measurements and point scanning measurements. For example, white light interferometry and the measurement method of structural illumination belong to wide field measurements, while the laser scanning confocal measurement belongs to point scanning measurements. For thin samples, the shadowing effect is negligible. Thus, the process of 3D imaging can be depicted accurately using the convolution of the optical system's PSF and object function. The coupling effect in optical measurements of the groove and step samples, can be demonstrated through the analysis of convolution.

A thin sample refers to the groove or step sample whose surface height difference is smaller than the axial main lobe of the optical system's PSF. Thin samples, can be illuminated evenly and the decoupling model of measurement can be established on the basis of the CIP. In other words, the decoupling area refers to the fact that there is no interference between the responses of points in the upper surface and those in the lower surface in the convolution in this area. Therefore the measurement data in this area is considered to be valid. The coupling area refers to the fact that there is interference between the responses of points in the upper surface and those in the lower surface in the convolution in this area. The measurement data in this area cannot be utilized effectively in a common method due to signal crosstalk, especially the unpredictable influence in superposition. It is regarded that the measuring data in this part suffers from a lack of fidelity.

Because the energy of the PSF is concentrated mainly on its main lobe and the lateral width is different from the axial width, the optical system PSF is simplified as an ellipse shape for convenience, as shown in figure 4.1. The system's response is the superposition of responses of every point in the sample. So when the ellipse symbolizing the PSF is at A_1 (as shown in the figure 4.1), the interference between the step edge and the lower surface of the step does not exist. When the ellipse is at A_2, the interference appears and the data of height measurement is not effective. When the ellipse is at the critical position A, interference is about to appear. In figure 4.1, a represents the axial radius of the PSF, b denotes lateral radius, and h is the step height.

As shown in figure 4.1, d is defined as the coupling distance (from point A to the edge). And when the distance between the area and step edge is less than d, this area is called the coupling area where measurement data is not valid. Taking the critical situation, it can be seen through analysis that when the step edge and the lower surface of the step just do not interfere with each other, the relationship between coupling distance and step height satisfies

$$\frac{d^2}{b^2} + \frac{h^2}{a^2} = 4.$$

(4.1)

The coupling distance of the lower surface can be obtained by analyzing the measuring process through the CIP. Similarly, the coupling distance of the upper surface also satisfies equation (4.1). Thus, when the step height of a thin sample is measured using an optical measurement, the avoidance distance d_{avoid} of effective portions from edges for height assessment should satisfy

$$d_{avoid} > d = b \times \sqrt{4 - h^2/a^2}.$$

(4.2)

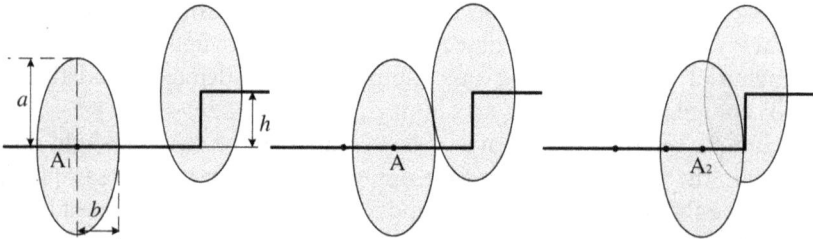

Figure 4.1. Illustration of the convolution irrelevance principle.

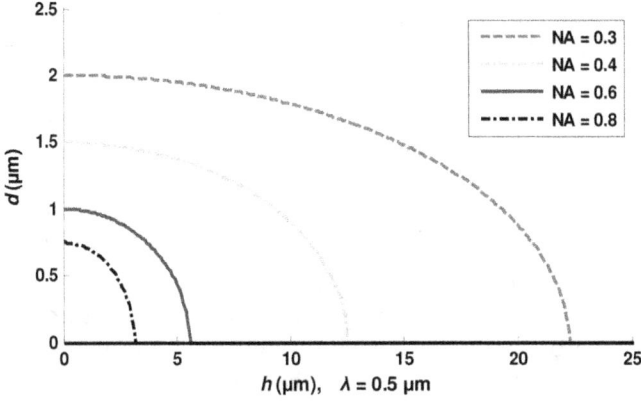

Figure 4.2. Simulation of the coupling relationships among d, h, and numerical aperture of a thin sample.

The ellipse radius of the PSF can be represented using the value of the first zero point of the main lobe. And the relationship between the decoupling distance of thin samples measured using an optical microscope and the step height of samples, where illumination wavelength λ and numerical aperture (NA), can be given by

$$d = \frac{0.3}{\text{NA}} \times \sqrt{16\lambda^2 - h^2 NA^4}. \tag{4.3}$$

As shown in figure 4.2 when the numerical aperture (NA) of an optical system is certain, the coupling distance of a thin sample and the step height have an inverse relationship. And when the step height of a thin sample is certain, the coupling distance of the thin sample and the NA of an optical system has an inverse relationship. The coupling distance is relevant to the PSF, thus the coupling distance is directly relevant to the lateral resolution while the thin sample is measured.

4.3 Decoupling model for the measurement of a deep groove sample

A deep groove sample refers to a sample whose surface height difference is greater than the optical system's PSF axial main lobe width. In this case, the illumination process and the collecting process will be obviously influenced by the shadowing effect. The shadowing effect will lead to an energy loss of illumination on the lower surface, and the amount of loss will relate to the height to width ratio of the groove. The actual light intensity received by the surface being measured has a significant influence on the signal light intensity, and the signal level determines the signal-to-noise ratio (SNR) of the system, thereby having an effect on measurement accuracy.

In this case, the decoupling area of the model is defined as the portion of the sample on which the optimal measurement accuracy can be achieved theoretically. The range of those areas can be determined with the limited energy lost (LEL) decoupling criterion. The specific requirement is that the percent of actual illumination light loss on those areas should not exceed 20%. The threshold value is selected based on the 0.8 Strehl ratio criterion. According to the criterion, if 80% of the total energy is received within an Airy disk, the imaging performance will not show significant degradation. Moreover, compared with decoupling areas, the area with actual illumination intensity loss of more than 20% suffers lower measurement accuracy.

Here the principle of the LEL decoupling criterion is illustrated with the example of confocal measurement. Optimal measurement accuracy can be guaranteed by ensuring the axial response signal has a high SNR and good symmetry (which are the keys to the axial location process) during the measurement. The signal level in the decoupling area can be guaranteed by keeping the energy of the illumination spot on the sample surface no less than 80% of that without spot defect and shadowing effect. Based on the LEL decoupling criterion, the reading rules for measuring a deep groove sample's height are illustrated in figure 4.3, in which D_d and D_s stand for the coupling range of the groove's upper and lower surfaces, respectively.

When the areas near edges are being measured, the reasons for the illumination light loss on the upper surface and the lower surface are different, as shown in figure 4.4. The light loss on the upper surface is due to the defect of the illumination spot while the energy loss on the lower surface can be attributed to the shadowing effect. Therefore, the formulas for calculating the coupling distances on the upper and lower surface are different.

Finally, the calculation formulas are as follows, and the detailed deduction can be found in [7]:

$$D_d = 1.2K\frac{\lambda}{\mathrm{NA}}, \tag{4.4}$$

$$D_s = K \cdot \begin{cases} 1.19\dfrac{\lambda}{\mathrm{NA}} + 1.49\dfrac{1 - \sqrt{1 - \mathrm{NA}^2}}{\mathrm{NA}}h, \ h \in \left(0, 0.97\dfrac{\lambda}{1 - \sqrt{1 - \mathrm{NA}^2}}\right) \\ 0.90\dfrac{\lambda}{\mathrm{NA}} + 1.79\dfrac{1 - \sqrt{1 - \mathrm{NA}^2}}{\mathrm{NA}}h, \ h \in \left(0.97\dfrac{\lambda}{1 - \sqrt{1 - \mathrm{NA}^2}}, +\infty\right). \end{cases} \tag{4.5}$$

Here, λ is the illumination wavelength; h is the step height of the deep groove sample; and K is a technological factor added to reflect how a real focusing differs from a theoretical one. The value of K is chosen according to the performance of the actual instrument because of the diversity of systems available, and usually greater than 1.

It can be seen from equation (4.4) that coupling range D_d of the upper surface is proportional to λ and inversely proportional to NA. The simulation of the

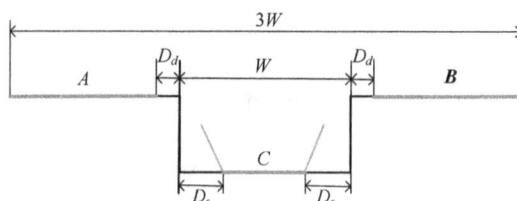

Figure 4.3. Limited energy lost (LEL) decoupling criterion.

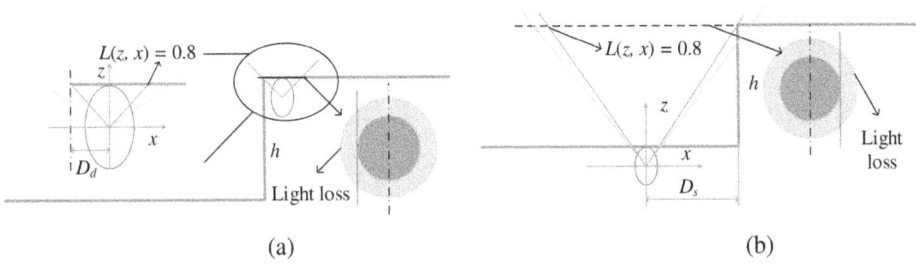

(a)   (b)

Figure 4.4. Illumination losses: (a) illumination loss for the step upper surface; (b) illumination loss for the step lower surface.

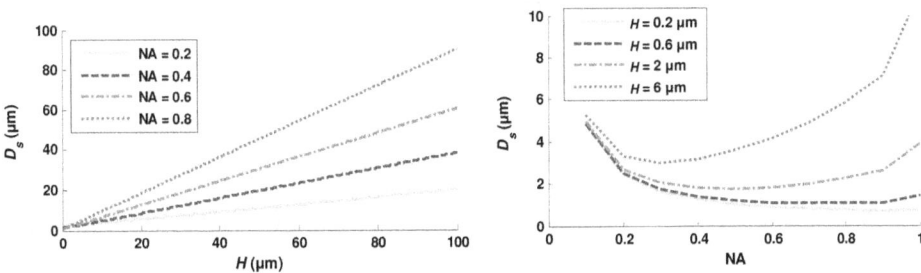

Figure 4.5. Simulation of coupling distance on the lower surface of groove samples ($\lambda = 0.405$ μm): (a) relationships between D_s and h with given NAs; (b) relationships between D_s and NA with given heights.

relationship between the coupling range D_s of the lower surface on the one hand and NA and h on the other hand is shown in figure 4.5.

It can be seen from figure 4.5(a) that D_s linearly grows as h increases with given λ and NAs. From figure 4.5(b), it shows that D_s monotonically decreases with the increase of NA when h is sufficiently small, and this tendency agrees well with the results from the analysis of thin samples in section 2.1. As h increases, a turning point appears when NA is large enough. If h belongs to a certain range, the D_s does not increase much as NA changes, which implies that when a larger NA objective is used the result will be better. However, when h is large, the shadowing effect becomes serious and dominates the influence, and the growing speed of D_s rapidly increases as h increases.

4.4 Experiments

A star pattern sample is a typical sample which is usually used for evaluating an optical instrument's lateral resolution. To verify the proposed decoupling criterion, star pattern sample ASP-0.2 R20 (provided by the National Physical Laboratory (NPL), UK) was measured using laser scanning confocal microscopy. According to the certificate of calibration from NPL, the maxim radius of the sample is 20 μm, the nominal measured mean height of the sample is 185.1 nm, and the expanded uncertainty of the measurement is 2.3 nm with a coverage factor of 2.0. In our measurement, the NA of the objective used is 0.8, and the laser wavelength is 532 nm. The measured topography is shown in figure 4.6(a).

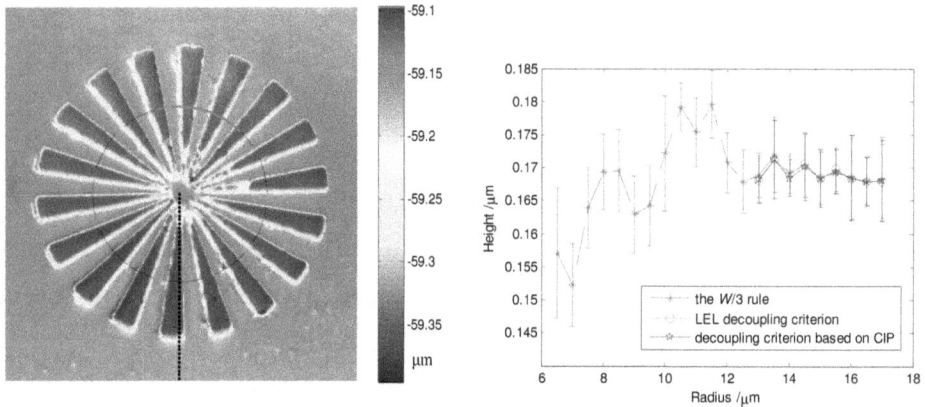

Figure 4.6. Measurements of a star pattern sample: (a) measured topology; (b) measured height of the ring section data along with the radius.

The ring section data of different radius can be taken as the result of grooves of different periods. According to equations (4.2), (4.3), and (4.4), the theoretical critical radius is 9.1 µm where the grooves' heights fail to be measured accurately. The critical radius, calculated based on the LEL decoupling criterion is 10.7 µm. There is a slight difference between the two results, due to the more restrictive conditions of the LEL decoupling criterion. Both criteria can be used to determine the distortion ranges of measurement results of thin samples. Moreover, the actual minimum measurable width of the grooves will be a little larger than the calculation, because the portions for height assessment cannot be infinitely small. Therefore, the actual critical radius will also be a little larger than the calculation.

From figure 4.6(b), it can be seen that: (1) the measured height values using the W/3 method obviously lack fidelity when the radius of the ring is small; and (2) the measured height values gradually converge to a steady value when the radius of the section ring is larger than 12 µm. This agrees well with theoretical predictions, and verifies that the 3D decoupling criteria can predict the ripple areas around the edge, thus indicating the effective portions for height assessment.

References

[1] Peng B, Giancarlo P and Wolfgang O 2012 Optical surface profile measurement using phase retrieval by tuning the illumination wavelength *Opt. Commun.* **285** 5029–36

[2] Mingkai Z *et al* 2015 Ultraviolet scanning linewidth measuring system *Infrared and Laser Engineering* **44** 625–31

[3] Meijing G *et al* 2013 Optical microscanning microscope thermal imaging system for electronic devices nondestructive testing *Laser & Infrared* **43** 779–84

[4] Liu J *et al* 2014 Digital differential confocal microscopy based on spatial shift transformation *J. Microsc.* **256** 126–32

[5] Martin R *et al* 2007 A landmark-based 3D calibration strategy for SPM *Meas. Sci. Technol.* **18** 404–14

[6] 2000 Geometrical product specifications (GPS)—Surface texture: Profile method; Measurement standards—Part 1: Material measures (Geneva: International Organization for Standardization).

[7] Jian L, Mengzhou L, Qiang L and Jiubin T 2016 Decoupling criterion based on limited energy loss condition for groove measurement using optical scanning microscopes *Meas. Sci. Technol.* **27** 125014 (12pp)

Chapter 5

Pupil filter design

In 2004, on the basis of a study into classic transfer function theory, B J Davis and others indicated that the consequence of adding superresolution filters in optical instruments to improve two point spatial resolution was a resultant cost of decreasing signal-to-noise ratio (SNR) of low-frequency information [1]. These filters improved the high-frequency signal's SNR and clearly presented some high-frequency information that was invisible without the filtering system. However, the cutoff frequency of the system's pass band was not extended. As a result, they believed that pure optical filters could not give the system the ability of superresolution in the strict sense.

In 2007, C J R Sheppard comprehensively expounded the nature of superresolution filtering, based on information theory [2]. The classic optical theory considered that the limited bandwidth of the transfer function reflected the fundamental limitation of a system's resolution ability. C J R Sheppard however believed that the necessary factors of a system, i.e. information-bandwidth product, cutoff frequency of pass band, signal period, field of view, polarization, SNR, and others, could reflect the fundamental limit of a system's resolution ability more effectively.

Based on whether information-bandwidth product was expanded or not, Sheppard judged an optical system to be a strict superresolution system. The so-called strict superresolution imaging system was the type based on encoding/decoding and the principle of time division multiplexing. The pass band could be expanded by adding observation time, polarizing the signal, modulating the wavelength or grating mask, multiplexing channels, and other ways, while superresolution imaging systems where the pass band was not extended, belonged in the non-strict category.

Sheppard encompassed the analyses and views of Davis and others regarding the nature of superresolution filtering technology which could improve spatial resolution. However, he further pointed out that in a confocal system, since the pinhole mask before the detector limited the radial field of view (FOV), the side lobe away from the focal center could be regarded as information outside the FOV. The side lobe could then be filtered by the pinhole mask, and in this way the resolution of the

doi:10.1088/978-1-6817-4337-0ch5   5-1   © Morgan & Claypool Publishers 2016 confocal system was improved. Superresolution effects combining confocal pinhole apodization with superresolution filters were defined as resolution improvements at the expense of limiting FOV. In a nonlinear response system, nonlinear effects of medium response produced side lobe apodization similar to a confocal pinhole, resulting in non-strict superresolution effects.

The main lobe width, side lobe magnitude, and Strehl ratio (SR), are three important parameters of the diffractive spot, and are used to evaluate the quality of the superresolution spot [3–5]. They also describe the level of superresolution, stray light suppression, and energy efficiency. In the aforementioned evaluation system of superresolution performance, three characteristic parameters of lateral superresolution, factor G, Strehl ratio S, and side lobe suppression ratio M, are normally used to represent lateral superresolution magnification, main lobe energy loss, and side lobe suppression level, respectively. Wherein, G is defined as the ratio of main lobes' first notch widths of superresolution and Airy spots, S is defined as the ratio of the main lobes' peak intensities of the above two spots, and M is defined as the ratio of peak intensities of the first lateral side lobe and the main lobe of the superresolution spot.

The side lobe is an important factor in damaging the quality of the diffractive spot [6, 7]. Inhibiting magnitude growth of the side lobe is one of the key issues in achieving superresolution confocal measurements. Superresolution filter design can be seen as a multi-parameter and multi-extremum global optimization problem to minimize the side lobe intensity, where the side lobe intensity is set as the objective function. Since the distribution of the side lobe intensity is a higher order structure function of the filter's parameters, and has a complex form of expression, existing design methods are problematic in ensuring adequacy and rigor of minimum optimization solutions in the entire complex defined interval [8, 9]. Seeking superresolution filter design methods which satisfy the necessary and adequate conditions of global minimum optimization for the side lobe is one of the important issues of current superresolution filter theory, requiring prompt resolution.

5.1 Phase rotation transformation

Side lobe suppression issues have become the main difficulty in superresolution filter design. On the one hand, since the position of the peak value of the side lobe changes with the filter's parameters, before completing filter design, it is difficult to accurately assess peak intensity of the side lobe. On the other hand, as the diffractive spot's transverse intensity distribution function belongs to a complex high order function, it is difficult to judge the sufficiency of global optimization for the resulting filter design. If the side lobe's amplitude is set as the objective function, as the function is complex, its extremum solutions cannot be obtained by applying existing extremum theory of real function. The function of phase rotation transformation is to transfer the complex amplitude of the side lobe into its modulus, thus ensuring the objective function becomes a low order real function in side lobe suppression design.

In a paraxial optical system, the focal spot's three-dimensional (3D) complex amplitude is calculated as

$$U(v, u) = 2\int_0^1 P(\rho)J_0(v\rho)\exp(iu\rho^2/2)\rho \, d\rho.$$

(5.1)

$P(\rho)$ is the pupil function, and ρ is the normalized radial coordinate of the annular pupil. $v = kr\text{NA}$ shows the focal spot's transverse dimensionless coordinate, where k, r, and NA refer to the wave vector, the radial coordinate in the focal plane, and the numerical aperture of the objective, respectively. $u = kz\text{NA}^2$ represents the spot's axial dimensionless coordinate, where z is the axial defocus amount.

Define $\mathbf{K} = [k_0 \ldots k_{N-1}]$, $\boldsymbol{\Phi} = [\varphi_0 \ldots \varphi_{N-1}]$, $j = 0, 1, \ldots, N - 1$. k_j and φ_j represent the transmittance and phase of the jth annulus, respectively. The complex amplitude of the annular pupil filter's 3D focal spot can be expressed as

$$U(v, u, \mathbf{K}, \boldsymbol{\Phi}) = 2\sum_{j=0}^{N-1} k_j e^{i\varphi_j} \int_{\rho_j}^{\rho_{j+1}} J_0(v\rho)\exp\left(iu\rho^2/2\right)\rho \, d\rho.$$

(5.2)

Affected by the phase modulation of the superresolution filter, the diffractive and geometric foci of the superresolution spot cannot perfectly coincide, and there is a slight offset u_F, where

$$u_F = -2\frac{\text{Im}\left(I_0^* I_1\right)}{\text{Re}\left(I_2^* I_0\right) - |I_1|^2}, \quad I_n = 2\sum_{j=0}^{N-1} \int_{\rho_j}^{\rho_{j+1}} k_j e^{i\varphi_j} \rho^{2n+1} \, d\rho.$$

(5.3)

To further illustrate the features of the diffractive spot's intensity distribution, figure 5.1 shows a contour graph of the 3D normalized intensity distribution of the diffractive spot without filter.

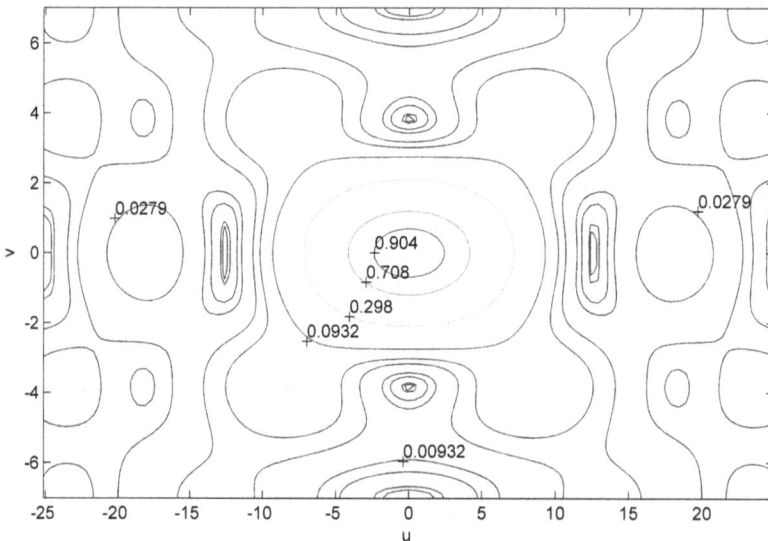

Figure 5.1. Contour graph of the 3D normalized intensity distribution of a diffractive spot.

Figure 5.1 shows that the 3D intensity distribution of the diffractive spot represents multi-peak and symmetric characters. When the offset between diffractive and geometric foci is ignored, transverse main lobe's amplitude reaches the maximum in the focal plane, meanwhile the first side lobe's amplitude is the highest compared with those of the others. As the defocus amount increases, the high order curve law ordered by the intensity of the horizontal main lobe has an obvious reduction, but the side lobe's attenuation is relatively slow.

There exists a noise enhancement phenomenon, which reveals that the intensity of the main lobe is below that of the side lobe, outside the range of axial focal depth. In addition to the neighborhood interval of the diffractive spot, the neighborhood range of lateral and axial zero points of the main lobe and side lobe is rich in high frequency signal components. Therefore simplifying the formula of the intensity of a 3D diffractive spot by using a low order series expansion method is unable to achieve accurate calculation results of the width and intensity of the main lobe and the magnitude of the side lobe intensity. Thus, employing low order series expansion methods to simplify the design of the objective function is bound to cause large calculation errors of intensity distribution outside the focal spot's neighborhood, therefore the scope of this method is limited. As a result, based on the precise complex amplitude calculation model of the diffractive spot, seeking a theory of simplifying the expression form of the objective function has a universal significance in solving the global optimization problem in superresolution filter design.

In terms of the superresolution diffractive spot, its intensity and the modulus of the complex amplitude can both reflect the magnitude of the superresolution spot's main lobe and the suppression feature of its side lobe. Considering the modulus of the spot's complex amplitude is a lower order function, setting the modulus as the objective function could therefore simplify the function's expression, enabling more convenient analyses and conclusions regarding the global properties of optimization solutions. Figure 5.2 shows the mathematical transformation idea of converting the complex function vector to the real function vector by phase rotation.

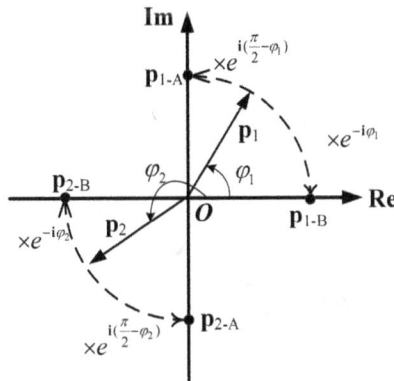

Figure 5.2. Schematic diagram of the phase rotating transformation.

Figure 5.2 shows that complex vectors $\mathbf{p}_1$ and $\mathbf{p}_2$ can be converted by the phase rotation transformation to pure virtual function vectors $\mathbf{p}_{1-A}$ and $\mathbf{p}_{2-A}$, or real function vectors $\mathbf{p}_{1-B}$ and $\mathbf{p}_{2-B}$, while the moduli of these vectors remain the same. This means that for any vector $\mathbf{p}$, there must exist a phase value φ which could make $\mathbf{p} \times \exp(-i\varphi)$ a pure virtual vector or real vector. The characteristic of keeping the moduli of vectors unchanged through the phase rotation transformation can be used to establish the objective function model in superresolution filter design with a simple form.

Choosing phase φ to satisfy $|U(v, u, \mathbf{K}, \mathbf{\Phi})| = |\mathrm{Re}[U(v, u, \mathbf{K}, \mathbf{\Phi}) \cdot \exp(-i\varphi)]|$, or $|U(v, u, \mathbf{K}, \mathbf{\Phi})| = |\mathrm{Im}[U(v, u, \mathbf{K}, \mathbf{\Phi}) \cdot \exp(-i\varphi)]|$, combined with equation (5.2) we can get

$$|U(v, u, \mathbf{K}, \mathbf{\Phi})| = \left| \mathrm{Re}\left[2 \sum_{j=0}^{N-1} k_j e^{i(\varphi_j - \varphi)} \int_{\rho_j}^{\rho_{j+1}} J_0(v\rho)\exp\left(iu\rho^2/2\right)\rho \, d\rho \right] \right| \qquad (5.4)$$

or

$$|U(v, u, \mathbf{K}, \mathbf{\Phi})| = \left| \mathrm{Im}\left[2 \sum_{j=0}^{N-1} k_j e^{i(\varphi_j - \varphi)} \int_{\rho_j}^{\rho_{j+1}} J_0(v\rho)\exp\left(iu\rho^2/2\right)\rho \, d\rho \right] \right| \qquad (5.5)$$

where Re and Im are arithmetic operators of real and imaginary parts, respectively. Since,

$$e^{i(\varphi_j - \phi)} = \cos(\varphi_j - \phi) + i \sin(\varphi_j - \phi)$$

when the diffractive focus is restricted to overlap with the geometric focus, i.e. $u_F = 0$, the following equations can be obtained:

$$|U(v, 0, \mathbf{K}, \mathbf{\Phi})| = \pm \sum_{j=0}^{N-1} k_j f_j(v, \rho) \sin\left(\varphi_j - \varphi\right) \qquad (5.6a)$$

$$\sum_{j=0}^{N-1} k_j f_j(v, \rho)\cos(\varphi_j - \varphi) = 0, \ u_F = 0 \qquad (5.6b)$$

or

$$|U(v, 0, \mathbf{K}, \mathbf{\Phi})| = \pm \sum_{j=0}^{N-1} k_j f_j(v, \rho) \cos\left(\varphi_j - \varphi\right) \qquad (5.7a)$$

$$\sum_{j=0}^{N-1} k_j f_j(v, \rho)\sin(\varphi_j - \varphi) = 0, \ u_F = 0 \tag{5.7b}$$

where

$$f_j(v, \rho) = 2 \int_{\rho_j}^{\rho_{j+1}} J_0(v\rho)\rho \, d\rho.$$

Equations (5.6) and (5.7) are equivalent, and equations (5.6b) and (5.7b) must exist as constraints for equations (5.6a) and (5.7a). Compared with the objective function based on the formula of intensity's distribution, the objective function based on equations (5.6a) or (5.7a) can be used to describe the amplitudes of a diffractive spot's main and side lobes, reducing the order of design variables of k_j and φ_j in the objective function.

In a non-paraxial optical system, since the polarization change of the focused beam cannot be ignored, the 3D spot intensity distribution model based on paraxial diffraction theory does not apply to focusing systems with large NA. In 1959, the following non-paraxial optical formula for calculating the complex amplitude in axisymmetric systems was put forward by Richards and Wolf:

$$U(r, z) = -\mathrm{i}kf \int_0^\alpha P(\theta)J_0(kr \sin \theta) \exp(\mathrm{i}kz \cos \theta) \sin \theta \sqrt{\cos \theta} \, d\theta \tag{5.8}$$

where $P(\theta)$ is a complex pupil function of the objective with large NA, and θ represents the angle between the optical axis and the propagation direction of a focused beam. Collecting angle α satisfied $\alpha = \arcsin(\mathrm{NA})$. The definitions of r and z are the same as those in equation (5.1). In terms of annular filter, the formula of complex amplitude of the diffractive spot can be further expressed as

$$U(r, z, \mathbf{K}, \mathbf{\Phi}) = -\mathrm{i}kf \sum_{j=0}^{N-1} k_j \mathrm{e}^{\mathrm{i}\varphi_j} \int_{\theta_j}^{\theta_{j+1}} J_0(kr \sin \theta)\exp(\mathrm{i}kz \cos \theta)\sin \theta \sqrt{\cos \theta} \, d\theta \tag{5.9}$$

where k_n and φ_n are defined the same as those in equation (5.2).

Repeating the analysis process in equations (5.6) and (5.7), the modulus of 3D amplitude of a diffractive spot with large NA can be obtained as

$$|U(r, 0, \mathbf{K}, \mathbf{\Phi})| = \pm \sum_{j=0}^{N-1} k_j f_j'(r, \rho)\cos(\varphi_j - \varphi) \tag{5.10a}$$

$$\sum_{j=0}^{N-1} k_j f_j'(r, \rho)\sin(\varphi_j - \varphi) = 0 \tag{5.10b}$$

or

$$|U(r, 0, \mathbf{K}, \mathbf{\Phi})| = \pm \sum_{j=0}^{N-1} k_j f_j'(r, \rho)\sin(\varphi_j - \varphi) \qquad (5.11a)$$

$$\sum_{j=0}^{N-1} k_j f_j'(r, \rho)\cos(\varphi_j - \varphi) = 0 \qquad (5.11b)$$

where

$$f_j' = kf \int_{\theta_j}^{\theta_{j+1}} J_0(kr \sin \theta)\sin \theta \sqrt{\cos \theta}\, d\theta.$$

Equations (5.10) and (5.11) are equivalent, and equations (5.10b) and (5.11b) must exist as constraints for equations (5.10a) and (5.11a). Equations (5.10) and (5.11) are lower order functions of design variables k_j and φ_j. It can be seen from the definition of multivariable function's differentiability, obtained by the phase rotation transformation and established under constraints, the modulus function of the diffractive spot's amplitude is a twice continuously differentiable function, whereby the convex property of which can be used to analyze the necessary and sufficient conditions for achieving the global optimization of the objective function. Based on a phase rotation transformation and under constraints, the formula used to calculate the modulus of the focal spot's complex amplitude provides a lower order mathematical model for study into the global optimization properties of objective functions in superresolution filter design.

5.2 The design method of filters with global minimizing side lobes

According to the extremum theory of a multi-variable function, if an objective function $f(x)$ is a twice-continuous differentiable convex function on the given convex set $\mathbf{D}$, and x^* serves as an inner point on $\mathbf{D}$, as well as $\nabla f(x^*) = 0$, then x^* is the solution of the global minimum of $f(x)$ within $\mathbf{D}$. The sufficient and necessary condition for $f(x)$ to be a convex function is that the Hessian matrix of $f(x)$ must be a positive definite or positive semidefinite matrix, which means the values of all-order principle minors are no less than zero, and the determinant values equal zero. If $H_f(x)$ is a positive definite matrix, $f(x)$ is a strict convex function, and x^* is the unique global minimum solution. The property of the convex function's extremum illustrates that if the objective function in superresolution filter design is convex, then the sufficiency of global optimization can be judged by the positive definiteness of the objective function's Hessian matrix.

Since the defined intervals of k_n and φ_n, [0, 1] and [0, 2π], belong to a convex set, if $|U(v, 0, \mathbf{K}, \mathbf{\Phi})|$ is a convex function, the local extremum solution of $|U(v, 0, \mathbf{K}, \mathbf{\Phi})|$ is the global extremum one. As equations (5.6) and (5.7) are equivalent to (5.10) and (5.11), equations (5.6) or (5.10) can be chosen as objective function in super-resolution filter design.

Suppose that the given $|U(v, 0, \mathbf{K}, \mathbf{\Phi})|$ in equation (5.6) is the objective function and the coordinates of peak value of the first side lobe serve as those of the maximum of the side lobe, then the Hessian matrix of $|U(v_M, 0, \mathbf{K}, \mathbf{\Phi})|$ can be calculated as

$$H_f(v_M, 0, \mathbf{K}, \mathbf{\Phi}) = \begin{bmatrix} \dfrac{\partial^2|U|}{\partial\mathbf{\Phi}^2} & \dfrac{\partial^2|U|}{\partial\mathbf{\Phi}\partial\mathbf{K}} \\[3mm] \dfrac{\partial^2|U|}{\partial\mathbf{K}\partial\mathbf{\Phi}} & \dfrac{\partial^2|U|}{\partial\mathbf{K}^2} \end{bmatrix} = \begin{bmatrix} c_{00} & \cdots & 0 & d_{00} & \cdots & 0 \\ \vdots & \ddots & \vdots & \vdots & \ddots & \vdots \\ 0 & \cdots & c_{N-1N-1} & 0 & \ddots & d_{N-1N-1} \\ d_{00} & \cdots & 0 & 0 & \cdots & 0 \\ \vdots & \ddots & \vdots & \vdots & \ddots & \vdots \\ 0 & \cdots & d_{N-1N-1} & 0 & \cdots & 0 \end{bmatrix}$$

where $c_{jj} = \mp k_j f_j(v_M, 0, \rho)\sin(\varphi_j - \phi)$, $d_{jj} = \pm f_j(v_M, 0, \rho)\cos(\varphi_j - \phi)$. In order to analyze the positive definite property of H_f easily, the above matrix is transferred into the following diagonal matrix:

$$\begin{bmatrix} c_{00} & \cdots & 0 & d_{00} & \cdots & 0 \\ \vdots & \ddots & \vdots & \vdots & \ddots & \vdots \\ 0 & \cdots & c_{N-1N-1} & 0 & \ddots & d_{N-1N-1} \\ d_{00} & \cdots & 0 & 0 & \cdots & 0 \\ \vdots & \ddots & \vdots & \vdots & \ddots & \vdots \\ 0 & \cdots & d_{N-1N-1} & 0 & \cdots & 0 \end{bmatrix} \xrightarrow{r_{N+j}+r_j\times\frac{-d_{jj}}{c_{jj}}} \begin{bmatrix} c_{00} & \cdots & 0 & d_{00} & \cdots & 0 \\ \vdots & \ddots & \vdots & \vdots & \ddots & \vdots \\ 0 & \cdots & c_{N-1N-1} & 0 & \ddots & d_{N-1N-1} \\ 0 & \cdots & 0 & \dfrac{-d^2{}_{00}}{c_{00}} & \cdots & 0 \\ \vdots & \ddots & \vdots & \vdots & \ddots & \vdots \\ 0 & \cdots & 0 & 0 & \cdots & \dfrac{-d^2{}_{N-1N-1}}{c_{N-1N-1}} \end{bmatrix}$$

where $j = 0, \ldots, N - 1$ and r_{N+j} and r_j are the $N+j$th and jth rows of matrix H_f, respectively. Then,

$$\prod_{j=0}^{p} c_{jj} \geqslant 0, \quad (-1)^p \prod_{j=0}^{p} d_{jj}^2 \cdot \prod_{j=p+1}^{N-1} c_{jj} \geqslant 0, \quad \forall p = 0, 1, \cdots, N - 1. \tag{5.12}$$

On the interval of $[0, 2\pi]$, since not all φ_j can satisfy equation (5.12), $|U(v_M, 0, \mathbf{K}, \mathbf{\Phi})|$ is not a convex function in the whole definition domain. It can be inferred that there must be non-unique separated convex subintervals in all of the intervals of objective function $|U(v_M, 0, \mathbf{K}, \mathbf{\Phi})|$, and inside these subintervals, local or global solutions for minimizing the side lobe in superresolution design could be obtained.

In order to make $|U(v_M, 0, \mathbf{K}, \mathbf{\Phi})|$ a convex function, and to make sure the local solution is the sufficient and necessary condition of global solution, a_j, b_j, and φ'_j could be defined as $a_j = k_j \cdot \sin\varphi'_j$, $b_j = k_j \cdot \cos\varphi'_j$, and $\varphi'_j = \varphi_j - \phi$, respectively. The objective function $|U(v_M, 0, \mathbf{K}, \mathbf{\Phi})|$ can be expressed as

$$|U(v_M, 0, \mathbf{K}, \boldsymbol{\Phi})| = \left| \sum_{j=0}^{N-1} a_j f_j(v_M, 0, \rho) \right| \tag{5.13a}$$

$$\sum_{j=0}^{N-1} b_j f_j(v_M, u_F, \rho) = 0, \, u_F = 0 \tag{5.13b}$$

where $a_j, b_j \in [-1, 1]$, $\varphi'_j \in [0, 2\pi]$, a_j and b_j are the design variables of equation (5.13).

From the definition of a differentiable function, since equation (5.13a) is a twice-continuous differentiable convex function and its Hessian matrix is zero, equation (5.13a) can be further proven to be a convex function inside the interval satisfying a_j, $b_j \in [-1, 1]$, and the extremum solution of equation (5.13a) is not unique. In this case, the local optimization solution is the global optimization one and thus the sufficient and necessary condition for a global optimization solution is automatically satisfied.

It can be seen from the above analyses that the design of an N-zone constant annular complex super-resolving filter (N-CACSF) equals the side lobe's global minimum problem described by

$$\min \left(\left| \sum_{j=0}^{N-1} a_j f_j(v_M, 0, \rho) \right| \right) \tag{5.14a}$$

subject to equation (5.6b), and the following constraints:

$$\left| U(v_g, 0, \mathbf{K}, \boldsymbol{\Phi}) \right| = 0 \tag{5.14b}$$

$$\frac{|U(0, 0, \mathbf{K}, \boldsymbol{\Phi})|^2}{I_A} = S \tag{5.14c}$$

$$\frac{\partial |U(v_M, 0, \mathbf{K}, \boldsymbol{\Phi})|}{\partial v_M} = 0 \tag{5.14d}$$

$$0 \leqslant a_j^2 + b_j^2 \leqslant 1 \tag{5.14e}$$

where

$$\frac{\partial |U(v_M, 0, \mathbf{K}, \boldsymbol{\Phi})|}{\partial v_M} = \mp 2 \sum_{j=0}^{N-1} a_j \int_{\rho_j}^{\rho_{j+1}} J_1(v_M \rho) \rho^2 \, d\rho \tag{5.14f}$$

$$k_j = \sqrt{a_j^2 + b_j^2} \tag{5.14g}$$

$$\varphi'_j = \tan^{-1}(a_j/b_j). \tag{5.14h}$$

In equation (5.14), $v_g = G \cdot v_{Airy}$ is the coordinate of the zero point of the superresolution spot's main lobe. v_{Airy} and I_A are the radius and the peak value of the Airy spot in the focal plane. k_j and φ'_j can be obtained by equations (5.14g) and (5.14h). It can be seen from equation (5.14) that the performance of a filter is decided by the difference in the phase value of annuli φ'_j, but not the absolute value of annuli φ_j, so ϕ given subjectively will not impact the performance of a filter, and φ_j can be replaced with φ'_j.

An initial value of v_M is needed for solving equation (5.14). The subjectively given initial value of v_M inevitably has its influence on the solutions of K* and **Φ***. However, v_M is related to the finalized parameters of the filter, so it is impossible to get accurate v_M before finalization of the parameters. To deal with v_M on a rational basis, a possible range of v_M can be established in accordance with the super-resolution law. v_M can be selected by searching for points in the possible range, and a set of filter parameters can be obtained by solving equation (5.14f) for the given v_M. The best set of filter parameters can then be selected through comparison. Equation (5.14f) is used to restrict the given v_M to be the coordinate of a stable point.

Compared with known design methods, the preceding developed method has the following advantages:

(1) By the phase rotation transformation and variable substitution, the modulus of the side lobe's amplitude is transferred into a convex function, and thus a design for optical filters with globally minimized side lobes is proposed. In the process of filter design, satisfying the sufficiency of global optimization in minimizing the side lobe ensures that the local extremum solutions of the objective function are the global extremum ones. Inside the full complex interval, the adequacy and tightness issues of annulus filter design with globally minimized side lobe could be solved.

(2) The developed method reveals the complex nature of optimization in the globally minimized side lobe design of superresolution filters, and it points out that the non-convex property of the objective function is the underlying cause that makes side lobe suppression complicated. The positive semi-definite feature of the Hessian matrix of the objective function being zero, further indicates that the solution of superresolution filter's global minimum is not unique.

(3) Although the above analyses are based on the design of N-CACSF, the annular width becomes narrower when N is large enough. An annulus with variable width can be expressed as a number of narrower constant annuli with the same phase value and the same width ($1/N$). The developed method is therefore applicable to the design of an annular filter with variable or constant width and a small or large number of annuli.

References

[1] Davis B J *et al* 2004 Capabilities and limitations of pupil-plane filters for superresolution and image enhancement *Opt. Exp.* **12** 4150–6
[2] Sheppard C J R 2007 Fundamentals of superresolution *Micron* **38** 165–9

[3] de Juana D M, Oti J E, Canales V F and Cagigal M P 2003 Design of superresolving continuous phase filters *Opt. Lett.* **28** 607–9

[4] Sales T R M and Michael Morris G 1997 Diffractive superresolution elements *J. Opt. Soc. Am. A* **14** 1637–46

[5] Sheppard C J R and Hegedus Z S 1988 Axial behavior of pupil-plane filters *J. Opt. Soc. Am. A* **5** 643–7

[6] Ledesma S *et al* 2004 Simple expressions for performance parameters of complex filters, with applications to super-Gaussian phase filters *J. Opt. Lett.* **29** 932–4

[7] Sheppard C J R 2007 Filter performance parameters for high-aperture focusing *Opt. Lett.* **32** 1653–5

[8] Liu C and Park S H 2004 Numerical analysis of an annular-aperture solid immersion lens *J. Opt. Lett.* **29** 1742–4

[9] Canales V F, Oti J E and Cagigal M P 2005 Three-dimensional control of the focal light intensity distribution by analytically designed phase masks *J. Opt. Commun.* **247** 11–8

IOP Concise Physics

Confocal Microscopy

Chenguang Liu, Jian Liu and Jiubin Tan

Chapter 6

Confocal axial peak extraction algorithm

6.1 Introduction

Topographic measurement of a three-dimensional (3D) surface is realized by determining the height of each point on the specimen according to the peak position of confocal axial response. For a metrological confocal microscope, the accuracy of the surface topographic measurement is directly affected by the axial peak position extraction algorithm. There are two aspects: the difference between axial position determinations; and the difference between axial position resolution and axial scale resolution. These need to be clarified before examination of the confocal peak extraction algorithm can begin.

The first misunderstanding is that the height of a specimen's surface by confocal microscopy is acquired by utilizing the linear region of the axial response envelope. This assumption confounds the principle of confocal microscopy with the traditional photoelectric sensing method. In fact, the topographic measurement of a micro-structure by confocal microscopy is realized by determining the position of peak value with fitting method. The characteristic of this procedure is that it can resolve a tiny step height, much smaller than axial scanning pitch. For example, typical sectional distances can be chosen in a commercial instrument in a range from 50 to 150 nm, while the height-position resolution is approximately 1–10 nm, which is much lower than the scanning pitch size. It is obviously different with the principle of photoelectric displacement sensors.

The second source of confusion is that the axial displacement resolution is misrecognized as the axial dimension resolution. Axial position resolution indicates the capability of detecting small displacements in axial direction while axial dimensional resolution refers to the ability to resolve two adjacent points located in the Z direction. They are mainly used in metrological measurement and 3D imaging of biological samples respectively. Axial position resolution depends on the accuracy of axial peak localization, while the axial dimensional resolution is determined by the half width of the axial response envelope.

6-1 © Morgan & Claypool Publishers 2016

6.2 Centroid method for localization of confocal peak

As a widely used confocal axial peak extraction algorithm, the centroid method is characterized by high processing speed and strong anti-noise ability. Supposing the 3D data obtained by confocal microscopy is $I(x_i, y_j, z_k)$, (x_i, y_j, z_k) is the position of scanning spots, the height of specimen surface is $h(x_i, y_j)$ which can be calculated using

$$h(x_i, y_j) = \Delta z \frac{\sum_{Z_k \in \text{FWHM}} I(x_i, y_j, z_k) z_k}{\sum_{Z_k \in \text{FWHM}} I(x_i, y_j, z_k)}. \tag{6.1}$$

Here, $Z_k \in$ FWHM indicates that only the data in the region near the center of the axial response is utilized to calculate the peak position, and its value is larger than half of the normalized peak value. Since the centroid method assigns the mass center of the axial response curve as peak position, when there is an aberration in the optical system (usually a spherical aberration) or the specimen has multi-reflecting layers, multi-peak appears in the axial response curve. In this case, the peak position determined by the centroid method has a large deviation, as shown in figure 6.1.

Z_r represents the actual peak value position of the confocal axial response. Z_c represents the peak value position calculated using the centroid method. It can be seen from figure 6.1 that there is a large deviation between these two results.

6.3 Nonlinear fitting method for peak localization

In order to overcome the shortcomings of the centroid algorithm, a fitting method based on a mathematical model is proposed to extract the position of axial peak, for example, a parabolic model or Gauss model.

The main problem of the current fitting method is that the approximate objective function is not consistent with the axial response of the confocal response theoretically. They only coincide appropriately in the neighbor region of the peak. If the fitting range of the axial response is enlarged, the number of fitting orders would increase dramatically, therefore lowering fitting efficiency and measurement

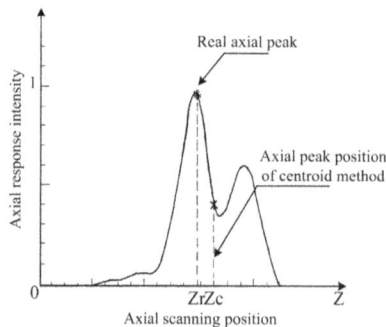

Figure 6.1. Deviation of the centriod method caused by the multi-peak in axial response.

speed. So it is of significance to analyze the confocal axial response model with different numerical apertures (NA).

I. Axial response model with small aperture.

According to coherent confocal imaging theory, the confocal imaging process can be described by equation (6.2)

$$I(x, y, z) = |h^2(x, y, z) \otimes u(x, y, z)|^2. \tag{6.2}$$

Here, $h(x, y, z)$ is the amplitude of the point spread function of a confocal system, $u(x, y, z)$ is an object function, and $\otimes$ represents a convolution operator. For a point-like object, the object function can be written in the form of $\delta(x, y, z)$. Through coordinate transformation,

$$\begin{cases} v = k\dfrac{a}{d_{10}}r \approx kr \sin \alpha_o \\ u = k\dfrac{a^2}{d_{10}^2}z \approx 4kz \sin^2(\alpha_o/2) \end{cases}, \tag{6.3}$$

where $r = \sqrt{x^2 + y^2}$, let '$v = 0$', equation (6.2) can be written as $I(u) = |h^2(0, u)|^2$, that is $I(u) = |h(0, u)|^4$. According to paraxial approximation, the normalized amplitude point spread function of a confocal system can be expressed as

$$h(v, u) = \int_0^1 P(\rho)J_0(v\rho)\exp\left(-ju\rho^2/2\right)\rho\,d\rho. \tag{6.4}$$

By substituting $I(u) = |h(0, u)|^4$ into equation (6.4)

$$I(u) = |h(0, u)|^4 = \left|\sin c\left(\frac{u}{4\pi}\right)\right|^4. \tag{6.6}$$

II. Confocal axial response model of plane object with a small aperture.

The object function of a plane object can be expressed as

$$t(v, u) = 1 \cdot \delta(u) \tag{6.7}$$

and the confocal axial response can be expressed as

$$I(u) = |h^2(v, u) \otimes \delta(u)|^2 = \left|\int_0^\infty h^2(v, u)v\,dv\right|^2. \tag{6.8}$$

The integral form of equation (6.8) with pupil function is

$$I(v, u) = \left|\int_{-\infty}^\infty \int_0^1 \int_0^1 P(\rho, u)J_0(v\rho)P(\rho', u)J_0(v\rho')\rho\rho'v\,d\rho\,d\rho'\,dv\right|^2. \tag{6.9}$$

Using the orthogonal characteristic of Bessel's function, the axial response of a confocal system can be simplified as

$$I(v, u) = \left| \frac{\sin(u/2)}{u/2} \right|^2. \tag{6.10}$$

III. Axial response model of a point-like object with a large aperture.

In the case of a small NA, the paraxial approximation can be used to simplify the complicated calculation. However, in order to realize higher resolution and accuracy in practice, especially for the measurement of submicron and even nanoscale structure, a large NA objective (NA > 0.5) should be used. In this case the condition for paraxial approximation is no longer satisfactory, and the effect of the apodization cannot be ignored. Firstly, utilizing the Debye approximation, the description of confocal focusing character can be expressed using the bellowing integral formula:

$$U(z) = -jkf \int_0^\alpha \exp(jkz \cos \theta)\sin \theta \sqrt{\cos \theta}\, d\theta. \tag{6.12}$$

By means of variable substitution, equation (6.12) can be written as

$$\begin{aligned} U(z) &= -jkf \int_{\cos\alpha}^1 \sqrt{\tau} \exp(jkz\tau)d\tau \\ &= -jkf \int_{\cos\alpha}^1 \sqrt{\tau}[\cos(kz\tau) + j \sin(kz\tau)]d\tau, \end{aligned} \tag{6.13}$$

where α is the aperture angle of an objective lens. $\sqrt{\tau}$ is not a typical function of the Fourier or cosine transform. It is difficult to simplify the integral to equation (6.13) as an analytic equation.

However, for a practical application, $\tau \in [\cos\alpha, 1]$, the range of τ is determined by the value of NA. Thus, there are two approximations established with $n \geqslant 2$ and $0 < c < d$ (where, n is an integer):

$$\begin{aligned} \int_c^d \tau^{1/n} \cos(kz\tau)d\tau &\approx \left[\frac{1}{d-c} \int_c^d \tau^{1/n}d\tau \right] \int_c^d \cos(kz\tau)d\tau \\ &= \frac{n\left(d^{\frac{1+n}{n}} - c^{\frac{1+n}{n}} \right)}{(1+n)(d-c)} \int_c^d \cos(kz\tau)d\tau \end{aligned} \tag{6.14}$$

$$\begin{aligned} \int_c^d \tau^{1/n} \sin(kz\tau)d\tau &\approx \left[\frac{1}{d-c} \int_c^d \tau^{1/n}d\tau \right] \int_c^d \sin(kz\tau)d\tau \\ &= \frac{n\left(d^{\frac{1+n}{n}} - c^{\frac{1+n}{n}} \right)}{(1+n)(d-c)} \int_c^d \sin(kz\tau)d\tau. \end{aligned} \tag{6.15}$$

Equations (6.14) and (6.15) can be verified through numerical simulation. Substituting these two equations into equation (6.12) gives

$$I(z) = |U(z)|^2 = A \sin c^2(z/B) \qquad (6.16)$$

where

$$\begin{cases} A = \dfrac{16\pi^2 f^2 \left(1 - \cos^{\frac{3}{2}} \alpha\right)^2}{9\lambda^2} \\ B = \dfrac{\lambda}{1 - \cos \alpha} \end{cases} \qquad (6.17)$$

In order to verify that equation (6.16) is consistent with equation (6.12), the curves of equations (6.16) and (6.12), and parabolic and Gauss functions, are plotted as shown in figure 6.2, with details shown in the sub-figure. As shown in figure 6.2, the point symbols slightly deviate from the blue line. The maximal deviation of the normalized intensity is 0.0058 at the main-lobe zero point and at the peak of the first side lobe. For the higher-order side lobes, the curves calculated using equations (6.16) and (6.12) virtually coincide. However, the curves calculated using the parabolic and Gaussian functions do not agree with the theoretically calculated intensity when the intensity decreases below 55% and 40% of the peak value.

IV. Model of flat axial response with a large NA.

The confocal axial response of planar object $\delta(z)$ with high NA can be expressed as

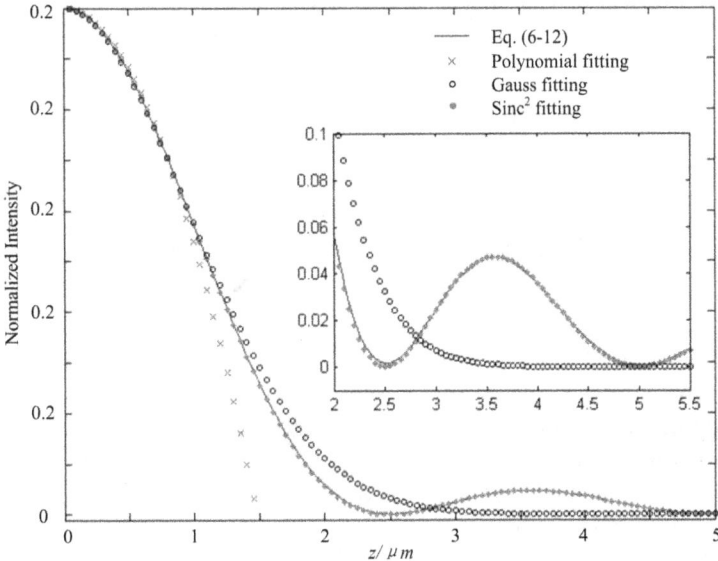

Figure 6.2. Fitting result of three models when NA is 0.8.

$$I(z) = |h^2(r, z) \otimes \delta(z)|^2 = \left| \int_0^\infty h^2(r, z)r \, dr \right|^2, \tag{6.18}$$

where r and z represent the radial and axial coordinate variables, respectively, in real space, $\otimes$ represents the convolution operator, and $h(r, z)$ represents the amplitude point spread function of an optical system with high NA.

According to equation (6.12), the integral of (6.18) can be written as

$$I(z) = \left| \frac{2\pi j}{\lambda} \int_0^\alpha \int_0^\alpha \int_0^\infty P(\theta)P(\theta')J_0(kr \sin \theta)J_0(kr \sin \theta') \right.$$
$$\left. \times \exp(-jkz \cos \theta)\exp(-jkz \cos \theta')\sin \theta \sin \theta'r \, d\theta d\theta' dr \right|^2, \tag{6.19}$$

where $J(\cdot)$ represents the zero-order Bessel function of the first kind. Based on the orthogonal property of the first Bessel function and the property of Dirac's function, equation (6.19) can be written as

$$I(z) = \left| \frac{2\pi j}{k^2\lambda} \int_0^\alpha P(\theta)^2 \exp(-j2kz \cos \theta)\tan \theta \, d\theta \right|^2. \tag{6.20}$$

Under a sinusoidal condition, apodization function $P(\theta)$ can be expressed as $\sin(\theta)$. Substituting it into equation (6.20) we can get the confocal axial response of a plane object, which can be expressed as

$$I(z) = \left| A \sin c\left(\frac{z}{B}\right) \right|^2, \tag{6.21}$$

where

$$\begin{cases} A = \dfrac{(1 - \cos \alpha)}{\pi k} \exp[-j(\cos \alpha + 1)kz] \\ B = \dfrac{\pi}{2k(1 - \cos \alpha)} \end{cases}. \tag{6.22}$$

6.4 Deviation analysis for localization of confocal axial peak

Due to the axial response envelope of incoherent confocal microscopy, there is no brief analytic expression. Deviations of several peak extraction methods are analyzed using the Monte-Carlo method. According to the principle of confocal microscopy profilometer surface measurement, the main factors having an effect on the accuracy of surface measurement are the axial scanning pitch of PZT and the additive Gauss noise of the detector. Their probability distributions are supposed to be the average distribution and the Gauss distribution, respectively.

Ideally, each scanning pitch of PZT can achieve the assigned position accurately. Supposing that the z-scanning pitch of PZT is z_{step} and the number of scanning sections is N, vector $\mathbf{Z} = (z_1, z_2, \ldots, z_i, \ldots, z_N)$ can represent the axial scanning position, where $z_i(i = 1, 2, \ldots, N)$ is the setting axial position of PZT. If the PZT initial position is 0, there is $z_i = (i - 1)z_{step}$. The signal strength corresponding to each scanning position can be calculated using $\mathbf{I} = f(\mathbf{Z}) = (I_1, I_2, \ldots, I_i, \ldots, I_N)$, $f(\cdot)$ represents the axial response function of incoherent fluorescence in confocal microscopy. The fitting process can be expressed as

$$\mathbf{A} = F[\mathbf{Z}, \mathbf{I}], \tag{6.23}$$

where $\mathbf{A}$ represents a vector set of parameters to be fitted, including the parameter denoting the peak position of confocal axial response, $F[\cdot]$ represents nonlinear fitting processing. The Levenberg–Marquardt algorithm is utilized as a nonlinear fitting algorithm to fit the axial response curve. However, in real measurement, the actual distance for each PZT step has a random deviation; that is, the PZT cannot be stabilized at set position $z_i = (i - 1)z_{step}$ exactly. Supposing PZT random deviations for step $\mathbf{\Delta Z} = (\Delta z_1, \Delta z_2, \ldots, \Delta z_N)$, which obey the uniform distribution and are independent from each other, the probability distribution function can be expressed as

$$g(\Delta z) = \begin{cases} \dfrac{1}{a}, & -\dfrac{a}{2} < \Delta z < \dfrac{a}{2}. \\ 0, & \text{else} \end{cases} \tag{6.24}$$

Therefore, the signal strength corresponding to each scanning position can be expressed as

$$\mathbf{I}' = f(\mathbf{Z}') = (I'_1, I'_2, \ldots, I'_i, \ldots, I'_N), \tag{6.25}$$

where $\mathbf{Z}' = \mathbf{Z} + \mathbf{\Delta Z} = (z'_1, z'_2, \ldots, z'_i, \ldots, z'_N)$.

At the same time, Gauss white noise is considered in this model, and its probability distribution can be expressed as

$$g(\varepsilon) = \frac{1}{\sqrt{2\pi}} e^{-\frac{\varepsilon^2}{2}}. \tag{6.26}$$

Then the actual fitting parameters can be written as

$$\begin{aligned} \mathbf{A}' &= F[\mathbf{Z}, \mathbf{I}' + \boldsymbol{\varepsilon}] \\ &= F[z_1, z_2, \ldots, z_i, \ldots, z_N; I'_1 + \varepsilon_1, I'_2 + \varepsilon_2, \ldots, I'_i + \varepsilon_i, \ldots, I'_N + \varepsilon_N]. \end{aligned} \tag{6.27}$$

Based on the model above, the $M = 10^4$ time's Monte-Carlo simulations are carried out with different axial sampling pitches using the centroid method, parabolic fitting method, Gauss fitting method, and $Sinc^2$ fitting method. The distributions of the results are shown in figures 6.3–6.5, and the standard standard deviations are listed in table 6.1. The uncertainty of the centroid method is larger than that of the fitting method, and it increases faster than the fitting method as the scanning pitch goes up. Moreover, the results show that there are few distributions in the region near $u = +0.3$, which is due to measurement error caused by the asymmetry of sampling

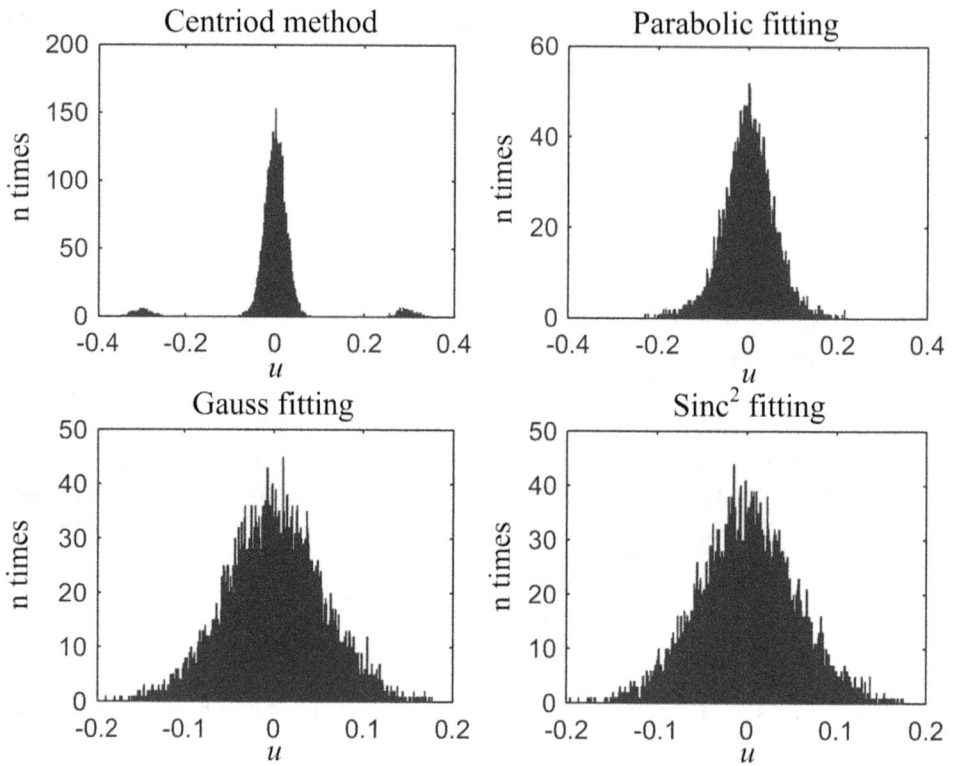

Figure 6.3. Probability distribution function showing the results of four peak extraction methods obtained using the Monte-Carlo method (pitch = 0.1).

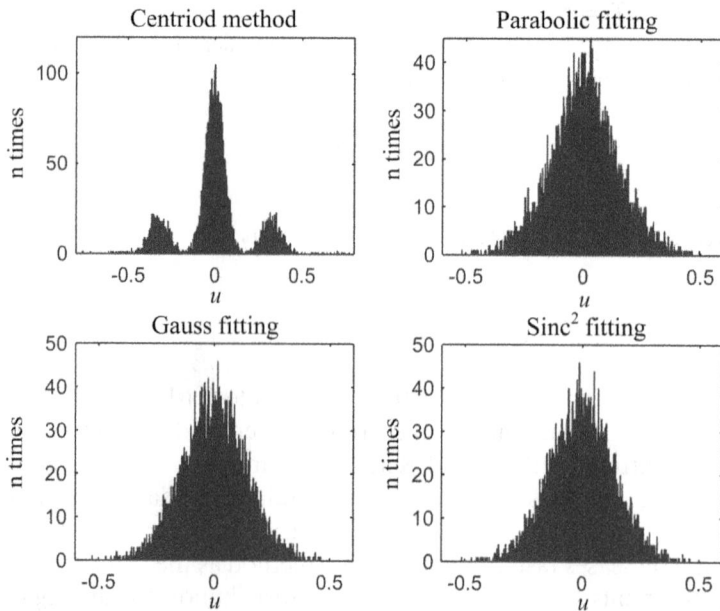

Figure 6.4. Probability distribution function showing the results of four peak value extraction methods obtained using the Monte-Carlo method (pitch = 0.5).

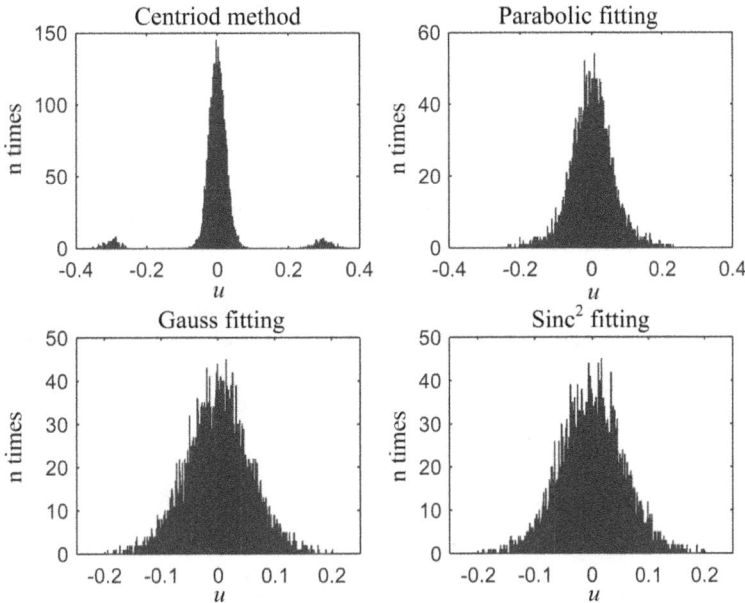

Figure 6.5. Probability distribution function showing the results of four peak extraction methods obtained using the Monte-Carlo method (pitch = 1.0).

Table 6.1. Position error of four methods under different steps (normalization optical coordinates).

Z pitch	Sinc^2 fitting	Gauss fitting	Parabolic fitting	Centroid method
0.1	0.0511	0.0516	0.0515	0.0691
0.5	0.0522	0.0526	0.0530	0.0711
1	0.1295	0.1297	0.1343	0.1778

points. With respect to fitting methods, the uncertainty of the Sinc^2 fitting model is lower than that of the Gauss model and the parabolic model. When the z-scanning pitch increases, the fitting uncertainty of the parabolic model increases faster than that of the Sinc^2 and Gauss models.

In this chapter, the theoretical model of coherent confocal axial response has been analyzed, and the Sinc^2 fitting method is proposed to achieve reliable peak value extraction. The accuracy of existing methods for peak localization in incoherent confocal microscopy are compared using the Monte-Carlo method, and a guiding reference id provided for the standard formulation of the confocal peak extraction algorithm.

References

[1] Tata B V R and Raj B 1998 Confocal laser scanning microscopy, Applications in material science and technology *Bull. Mater. Sci.* **21**(4) 263–78

[2] Udupa G *et al* 2000 Characterization of surface topography by confocal microscopy, I. Principles and the measurement system *Meas. Sci. Technol.* **11**(3) 305

[3] Li S G, Xu Z G, Reading I, Yoon S F, Fang Z R and Zhao J 2008 Three dimensional sidewall measurements by laser fluorescent confocal microscopy *Opt. Express* **16** 4001–14

[4] Köklü F H, Quesnel J I and Vamivakas A N 2008 Widefield subsurface microscopy of integrated circuits *Opt. Express* **16** 9501–6

[5] Dumitriu D, Rodriguez A and Morrison J H 2011 High-throughput, detailed, cell-specific neuroanatomy of dendritic spines using microinjection and confocal microscopy *Nat. Protoc.* **6** 1391-411

[6] Liu J, Tan J, Liu T, Wan S and Cao Ya 2010 Axial performance parameters developed for analytical design of center shaded filter in high aperture focusing system *Opt. Commun.* **283** 4190–3

[7] Tan J *et al* 2015 Sinc^2 fitting for height extraction in confocal scanning *Meas. Sci. Technol.* **27** 025006

Chapter 7

Differential confocal microscopy

7.1 Introduction

The axial position of surface measurement is assessed and obtained by locating the axial responding curve in a confocal microscopy system with high precision and thereby eliminating the problems of nonlinear error that exist in the displacement sensor.

However, in order to get high measurement efficiency or an improved capability of surface tracking and detecting in real situations, the system has to take the surface measurement without acquiring all the information from the axial responding curve. The problem can be summarized down to the tomographic measurement of the system not possessing the required bipolar tracking characteristics. Thus, confocal microscopy cannot assess the direction of defocus or fast track measurement in real time.

With this problem in mind, a differential confocal microscopic method was proposed by the Harbin Institute of Technology (HIT) in 2000 [1]. The differential confocal microscopic concept can be traced back to 1982 when it was proposed by the Israeli scholars, Yeshaiahu Fainman, Ehud Lenz, and Joseph Shamir [2]. They acquired a high axial response sensitivity at the 0.1 μm level, a spatial resolution at the 2 μm level, and expanded the measurement range to the millimeter scale by introducing positive and negative-going defocus differential detectors in order to achieve their differential confocal microscopy (DCM). In 1994, Andrew W Kulawiec and Duncan T Moore from the University of Rochester achieved phase quantization imaging using the method of differential confocal detection [3]. The differential confocal microscopic technology proposed by HIT differs from the aforementioned method, both in the research aims and in the new characteristics sought. This proposal for confocal microscopy technology sought to acquire the bipolar tracking characteristic. Fast tracking of surface measurement can be realized by assessing the zero-crossing of a differential responding curve.

doi:10.1088/978-1-6817-4337-0ch7
7-1
© Morgan & Claypool Publishers 2016

The similarity of differential confocal microscopic technology and traditional confocal microscopy technology is that they both realize the tomographic measurement by the assessment of the characteristic point. Therefore, neither of them has the unavoidable nonlinear error in the linear displacement sensor. However, the former has a better real time tracking characteristic and is also suitable for dynamic tracking scanning in a wide range. In three-dimensional (3D) measurement, the way to acquire the accurate location of the surface measurement is to detect the zero-crossing of the differential response curve by fitting the axial differential curve.

7.2 Application of DCM in China

In 2002, Jiubin Tan and Fusheng Wang of HIT proposed DCM to fabricate a differential confocal sensor with high sensitivity and high linearity in order to overcome the shortcomings that the trigger feature point of the optical probe has in the low linearity area, which cause a low distinguishability of axial direction [4]. Therefore, 3D microstructural samples can be measured with high axial resolution.

In 2003, the gradient-index differential confocal microscopy (GIDCM) was proposed by Jiubin Tan and Jie Zhang which has quantitative, miniaturization, and easy-coupling advantages of the gradual index lens compared with traditional microscopy [5]. As a result, the new characteristic of on-line measurement can be achieved in a narrow space using a midget differential confocal sensor based on GIDCM.

In 2004, bipolar absolute differential microscopy (BADCM) was proposed by Weiqian Zhao, Jiubin Tan and others, in order to achieve the capability of high axial resolution and transverse superresolution imaging in a measurement system, by introducing a superresolution filter in the differential confocal optical path [6–8]. In the same year, tri-heterodyne confocal microscopy (TCM) was proposed by the same researchers, in which the new sensing characteristic curve has the ability of axial superresolution, using TCM with the help of three point detectors located at the focus, positive-going defocus, and negative-going defocus plane [9, 10].

In 2009, improved differential confocal microscopy (IDCM) was proposed by Jian Liu and Jiubin Tan in order to rebuild the sensing characteristic curve with the new characteristics of high signal-to-noise ratio and disturbance rejection resistance of the system, which uses the difference determined by the sum of the two signals from the two detectors [11].

In the same year, phase-shifting interference microscopy was proposed and developed to improve measurement resolution and avoid phase unwrapping. In 2010, differential confocal microscopy based on polychromatic illumination (DCMBPI) was also proposed by these researchers, which acquires the doubled axial measured linear region in comparison with traditional differential confocal microscopy [12]. By designing the combination of two different differential confocal sensing characteristic curves produced by two light sources in different wavelengths, this method overcomes the traditional differential confocal microscopy's disadvantage of short limitation in height measurement.

In 2010, dual-axed differential confocal microscopy (DaDCM) was proposed by Weiqian Zhao, Qin Jiang, and others [13]. In this method, the incident illumination light path and reflection detection light path of differential confocal systems are symmetrical in the normal line. The two virtual pinholes are symmetrical on the CCD left and right sides respectively, to be a positive-going defocus differential detector and negative-going defocus differential detector, in order to acquire the differential confocal sensing curve and to complete the measurement. This method is suitable for high axial resolution measurement with the limitation of long working distance. In the intervening period, two virtual point detectors have been constructed in one CCD imaging plane, which has overcome the problem of strictly symmetrical installations for the two point detectors in traditional differential confocal microscopy.

In 2012, laser divided-aperture differential confocal sensing technology (LDDCST) was proposed by Weiqian Zhao, Chao Liu, and others [14]. This technique produces positive-going and negative-going defocusing laser spot separating, symmetrically relative to the optical axis center, by dividing the microscopic objective pupil into two tangential subapertures. It obtains positive-going and negative-going defocusing differential signals by constructing two virtual pinholes with a certain group of CCD pixels. The two signals then construct a differential confocal sensing characteristic curve in order to solve the limitation of an unusable high numerical aperture of the objective lens and improve the system's spatial resolution.

In 2012, radial birefringent pupil filtering differential confocal microscopy (RBPFDCM) was proposed by Limin Zou and others [15, 16]. This method improves the measurement system spatial resolution by introducing radial birefringence pupil filtering to differential confocal systems.

In 2013, lateral differential confocal microscopy (LDCM) was proposed by Cuifang Kuang and others [17, 18]. This method comprises the moving of optical fibres which are located on the focus plane, laterally and symmetrically. It has a new feature of accurately positioning the profile of a step edge under low signal-to-noise ratio conditions.

In 2014, digital differential confocal microscopy (DDCM) was proposed by Jian Liu and Jiubin Tan, which highlights the fact that the differential confocal is a kind of application of the axial spatial shift transformation invariance characteristic in nature, and the DDCM can realize this characteristic by digital processing [19]. Thus, the DDCM has the characteristic of programmable axial detector displacement, in order to dynamically solve the problems of axial detector displacement matching when switching to different objectives. It is also able to exempt the precision installation requirement, reduce equipment costs, and increase reliability.

7.3 The Basic principle of DDCM

Differential confocal microscopy is a type of technology which has the characteristic that the axial installation error of a detector only produces shift movement of the axial responding curve, without changing its 3D spatial invariance in order to realize common-mode noise suppression by multi-detector differential confocal detection. The method of locating the sample's Z position is changed from assessing the axial response peak to detecting the zero-crossing of the differential responding curve. The advantage of the differential detection concept is its ability to suppress common-mode noise and increase axial response sensitivity. However, it should be noted that traditional DCM reduces spatial resolution while increasing axial response sensitivity, therefore, the use of DCM should be based on this specific precondition.

In fact, the reason for the reduction of spatial resolution is not difficult to understand using the basic principle of DCM. As DCM is based on multi-detector hardware, the optimum lateral focusing position of different channels is divided at the same time, when various channel signals are shifted by axial displacement under the researcher's direction, even if the optimum lateral focusing plane is separate from the optimum axial focusing position of the two detectors. This means the system cannot reach its optimum 3D resolution position at the confocal location of DCM theory, but will reduce with the increase of differential detector sensitivity.

In the introduction of DCM technology, the core concept of DCM was to take advantage of the feature that the axial shift installation of the detectors would not influence its 3D spatial invariance but only produce the axial shift transformation on axial direction. Therefore, it is clear that DCM technology is a kind of shift transformation in terms of mathematics, thus DDCM technology has been proposed by HIT researchers, and the working principle is shown in figure 7.1.

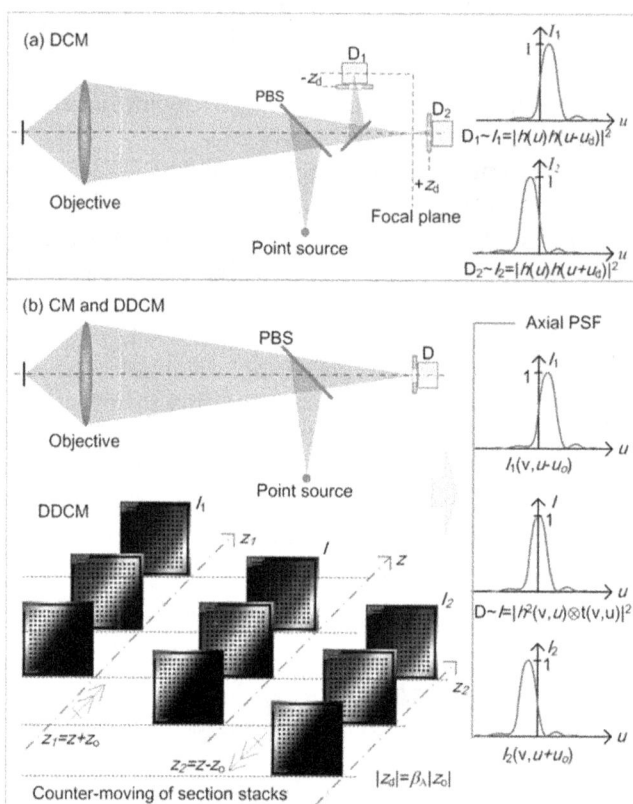

Figure 7.1. Working principle of a DDCM. Reproduced with permission from Liu J, Wang Y, Liu C, Wilson T, Wang H, and Tan H 2015 *J. Microsc.* **256** 126. Copyright 2014 John Wiley and Sons.

Shown in figure 7.1(a), two divided detectors are used in a traditional differential confocal system, the efficient point spread functions are $h(v, u)h(v, u \pm u_d)$, and the defocus displacement of the detector is represented by u_d whose value can control the separation distance of two confocal axial response curves. The installation error and the asymmetry will reduce the linear range and axial response sensitivity. From the illuminated and detection arm ($h(v, u)$ and $h(v, u \pm u_d)$) of the differential confocal system in the theory model, the best illumination position and detection position locate respectively at $u = 0$ and $u = \pm u_d$, while the optimum sensitivity position of differential confocal output is at $u \approx u_d/2$. The lateral resolution is in the defocus condition when the system reaches the best axial resolution, which means the 3D resolution will decrease. This is the original defect of DCM, based on the hardware separation and detection principle. However, DDCM is different from the principle above. The system response function is shown in figure 7.1(b) and stated as

$$I(v_s, u_s, \pm u_0) = |h^2(v_s, u_s \pm u_0) \otimes t(v_s, u_s)|^2, \tag{7.1}$$

in which $t(v_s, u_s)$ is the objective function while v_s and u_s are the transverse and the axial optical coordinates respectively, $\otimes$ is the convolution operator, and u_0 is the digital axial detector displacement caused by the stack movement. u_0 is a programmable parameter which is calculated by $n \times ds$.

It is necessary to ensure the optimum axial defocusing condition. We calculate the shift number n of transformation sections, equivalent defocusing distance of the object space z_o, and the scanning step d_s according to the analysis of DCM's response characteristic before executing the numerical calculation of DDCM. In theory, the parameters can be changed flexibly in DDCM to match different numerical aperture parameters and ensure the optimum axial detection sensitivity can be acquired in different measurement conditions. As to the scanning step, the d_s value is usually selected from 30 nm to 150 nm. The suitable value of scanning step is conducive to increasing the scanning speed, however, an excessive scanning step will decrease the fitting precision. Moreover, if the scanning step is undersized, the scanning time and data size will definitely grow with no increase in precision. Thus, the problem of choosing the optimum scanning length needs to be carefully considered. The comparison explanation of the principle, for traditional DCM and DDCM, is shown in the following.

As shown in figure 7.1, $u = 8\pi\sin^2(\alpha/2)z/\lambda$ is the normalized axial optical coordinate, α is defined as objective semiaperture angle, and u_o and u_d are the axial optical coordinates of the object space and detector space. The corresponding coordinate of u_o is u_d, and β_A is the axial magnification. In actual application of DDCM, $z_o = n \times d_s$, n is the shift number n of transformation sections and d_s is the axial scanning step. The pinhole detectors are represented by D and $D_{1,2}$. The specific expression of $h(v, u)$ in the point spread function is

$$h(v, u) = \int_0^\alpha J_0\left(\frac{v \sin \theta}{\sin \alpha}\right) e^{iu\sin^2(\theta/2)/2\sin^2(\alpha/2)}\sqrt{\cos \theta} \sin \theta \, d\theta, \tag{7.2}$$

$I(v_s, u_s, +u_0) - I(v_s, u_s, -u_0)$ is used to expresses the finial output of DDCM. The axial location of measurement points can be judged by the zero-crossing extract of the differential responding curve. As technology based on the curve fitting algorithm has been sufficiently expounded and applied in different kinds of engineering technologies, the introduction of the algorithm is therefore omitted from this chapter. The developers and researchers of equipment can select fitting algorithms which are more flexible and faster in order to increase the extraction accuracy and calculation speed according to requirements in real circumstances.

From figure 7.2, when defocus displacement increases from 3 to 6 in optical dimensionless coordinates, the strength scope of the side lobe increases from 0.027 to

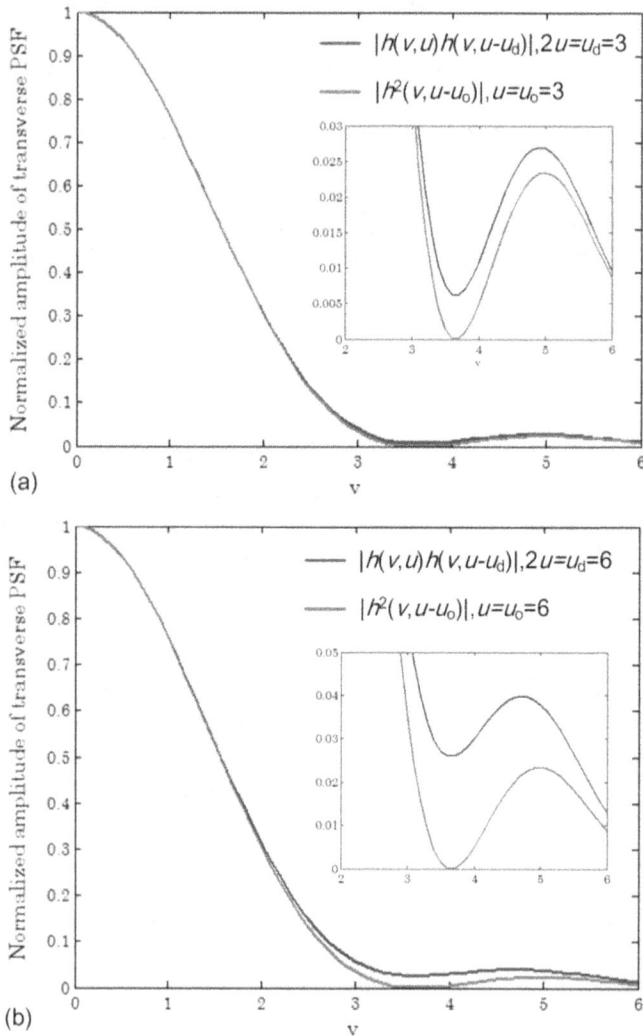

(a)

(b)

Figure 7.2. Simulation of the relationship between point spread function and defocus displacements of differential point detectors. Reproduced with permission from Liu J, Wang Y, Liu C, Wilson T, Wang H, and Tan H 2015 *J. Microsc.* **256** 126. Copyright 2014 John Wiley and Sons.

0.04, with an increase of 48%. At the same time, full width at half maximum (FWHM) is also broadened accordingly.

According to optical theory, an increase of the side lobe will lead to a rapid decline in the contrast of medium-frequency and low-frequency information in the transfer function, which shows a decrease of signal-to-noise ratio. For measurement equipment, all these factors will directly influence the accuracy of results.

For the technology of DDCM, every frame of scanning images is acquired by the basic confocal microscopy model, which is in optimum 3D spatial resolution. Therefore, compared with DCM, there are no such problems, like position separation of acquiring best resolution in axial and lateral, and an increase of side lobe because of defocusing detectors.

Certainly, only by differential calculations can a system decrease the common-mode noise. At the same time, differential output will introduce differential detection responses. However, this digital detection model improves fundamentally the conventional acquirement of data for DCM and overcomes the theoretical disadvantage of reduction of image resolution in the process of obtaining raw data. In addition, the theory of DDCM explains the essence of differential technology; that is, 'spatial shift transformation invariance'.

Furthermore, in specific applications and conditions, further development will be easily achieved based on DDCM, such as multiple differential confocal detection, lateral differential confocal detection, and other detection methods. Meanwhile the developed instrument can dispense with the complex installation of detectors with strict symmetry, decrease costs, and improve reliability.

References

[1] Wang F S, Tan J B and Zhao W Q 2000 *The optical probe using confocal technique for surface profile measurement* C SPIE 4222 194–7

[2] Fainman Y, Lenz E and Shamir J 1982 Optical profilometer: a new method for high sensitivity and wide dynamic range *Applied Optics* **21** 3200–8

[3] Kulawiec A W and Moore D T 1994 *Quantitative phase imaging in confocal microscopy by optical differentiation* **33** 6582–90

[4] Tan J and Wang F 2002 Theoretical analysis and property study of optical focus detection based on differential confocal microscopy *J. Meas. Sci. Technol.* **13** 1289

[5] Tan J and Zhang J 2003 Differential confocal optical system using gradient-index lenses *J. Opt. Eng.* **42** 2868–71

[6] Weiqian Z, Jiubin T and Lirong Q 2004 Bipolar absolute differential confocal approach to higher spatial resolution *J. Optics Express* **12** 5013–21

[7] Weiqian Z, Jiubin T and Lirong Q 2005 Improvement of confocal microscope performance by shaped annular beam and heterodyne confocal techniques *J. Optik* **116** 111–7

[8] Weiqian Z *et al* 2005 SABDCM a new approach to higher lateral resolution of laser probe measurement *J. Sensor and Actuators* 17–25

[9] Weiqian Z, Jiubin T and Lirong Q 2004 Tri-heterodyne confocal microscope with axial superresolution and higher SNR *J. Optics Express* **12** 5191–7

[10] Weiqian Z, Zhengde F and Lirong Q 2007 A shaped annular beam tri-heterodyne confocal microscope with good anti-environmental interference capability *J. Chin. Phys.* **16** 1624–31

[11] Jian L *et al* 2009 Improved differential confocal microscopy with ultrahigh signal-to-noise ratio and reflectance disturbance resistibility *J. Appl. Opt.* **48** 6195–201

[12] Tan J, Liu J and Wang Y 2010 Differential confocal microscopy with a wide measuring range based on polychromatic illumination *J. Meas. Sci. Technol.* **21** 054013

[13] Zhao W *et al* 2011 Dual-axes differential confocal microscopy with high axial resolution and long working distance *J. Opt. Commun.* **284** 15–9

[14] Zhao W, Liu C and Qiu L 2012 Laser divided-aperture differential confocal sensing technology with improved axial resolution *J. Optics express* **20** 25979–89

[15] Limin Z *et al* 2012 Differential confocal technology based on radial birefringent pupil filtering principle *J. Opt. Commun.* **285** 2022–7

[16] Zou L *et al* 2013 Response characteristics of differential confocal system based on radial birefringent pupil *J. Opt. Commun.* **303** 15–20

[17] Wang Y *et al* 2014 A lateral differential confocal microscopy for accurate detection and localization of edge contours *J. Opt. Lasers Eng.* **53** 12–8

[18] Wang Y *et al* 2013 Image subtraction method for improving lateral resolution and SNR in confocal microscopy *J. Optics & Laser Technology* **48** 489–94

[19] Liu J, Wang Y, Liu C, Wilson T, Wang H and Tan J 2014 Digital differential confocal microscopy based on spatial shift transformation *J. Microsc.* **256** 126–32

IOP Concise Physics

Confocal Microscopy

Chenguang Liu, Jian Liu and Jiubin Tan

Chapter 8

Medium aided scattering measurement

8.1 Introduction

The limited aperture of an optical system not only restricts the imaging and measurement resolution, but also the detection capability of smooth, highly curved, or tilted surfaces. Even when such surfaces are imaged using high aperture microscope objectives, the surfaces of steep features cause the light to be reflected in such a way that it is not captured by the lens. This is true even in the limiting case of a unity numerical aperture, since the illuminating light may also be reflected in the forward direction. This principle limitation cannot be overcome, even by using high numerical aperture. The imaging of smooth, highly curved, or tilted surfaces is widely recognized as one of the most challenging and unsolved problems in optical imaging and metrology today.

8.2 The principle of medium aided scattering confocal microscopy

Medium aided scattering confocal microscopy has been proposed in order to accomplish the shape measurement of smooth, highly curved, or tilted surfaces. An easily removable fluorescent layer with nanoscale thickness is deposited on the specimen by means of physical vaporization deposition (PVD) to generate an isotropy fluorescent scattering. The scattering fluorescent light with surface location information will always fill the aperture of the optical system; and the radiation centre of fluorescent light does not change with the inclination angle of the specimen surface. The principle is shown in figure 8.1(c).

One of the critical factors is that the fluorescent medium layer can be easily removed without damage to, or pollution of, the specimen. Rhodamine B, a type of organic fluorescent material with low sublimation temperature, will not crystallize

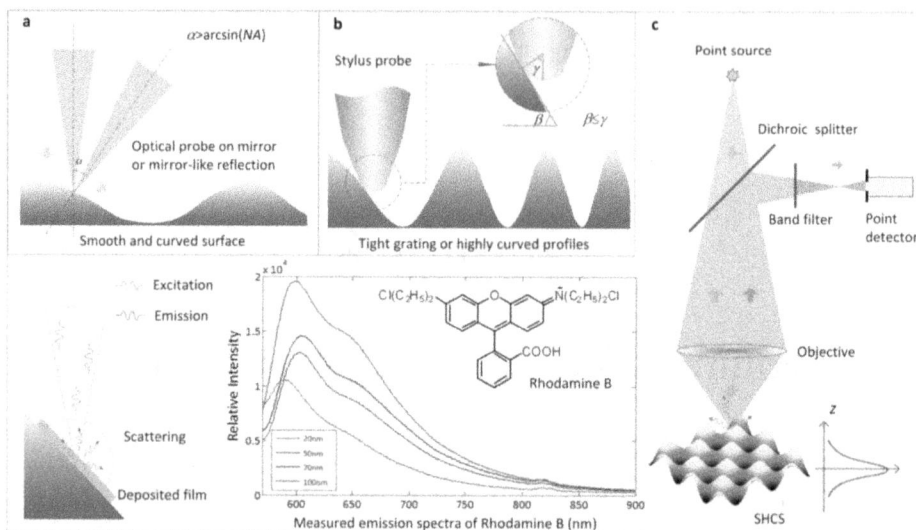

Figure 8.1. Principle of medium aided scattering confocal microscopy. (a) The optical problem; reflections are not collected by the objective lens. Here α is the angle between normal incidence and the reflected beam, NA denotes the numerical aperture of the objective lens, and light is not collected if $\alpha > \sin^{-1}(NA)$. (b) The stylus problem; β, the maximum local slope, must be less than γ to reveal a true profile. (c) The confocal fluorescence microscope and the measured excitation and emission spectra of the rhodamine B used in this study. The filter is selected to pass wavelengths greater than 580 nm in order to separate the excitation and emission light. The peak of the emission spectrum is around 600 nm with a 552 nm excitation peak. Reproduced with permission from Liu J, Liu C, Tan J, Yang B, and Wilson T 2015 *J. Microsc.* **261** 300. Copyright 2015 John Wiley and Sons.

on the surface of a sample during the PVD process. Furthermore, it is highly soluble in water and solutions of hydrochloric acid, ethanol, chloroform, and sodium hydroxide, and can be easily removed ultrasonically.

8.3 Analysis of deposition uniformity of a fluorescent medium layer

In order to analyze the thickness variation of a medium layer with an inclination angle of the specimen surface, the deposition thickness of rhodamine B on quartz sheets with various inclination angles is compared. The inclination angle is set from 0° to 90° with an increment of 10°, and the deposition thickness is measured by the Alpha Step D-100 profile-system, whose height resolution is 0.6 nm.

The deposition thickness is set to 150 nm, which is monitored and controlled by a deposition thickness monitor (si-tm206c) within the PVD device. The results are shown in figure 8.2. The thickness decreases from 150 nm to 30 nm with the inclination angle changing from 0° to 90°, and the variation of thickness is negligible when the inclination angle changes from 0° to 30°. This result indicates the deposition thickness variation in extreme cases and the necessity to optimize the deposition process to correct height error caused by a non-uniform layer when measuring a surface with a large inclination angle.

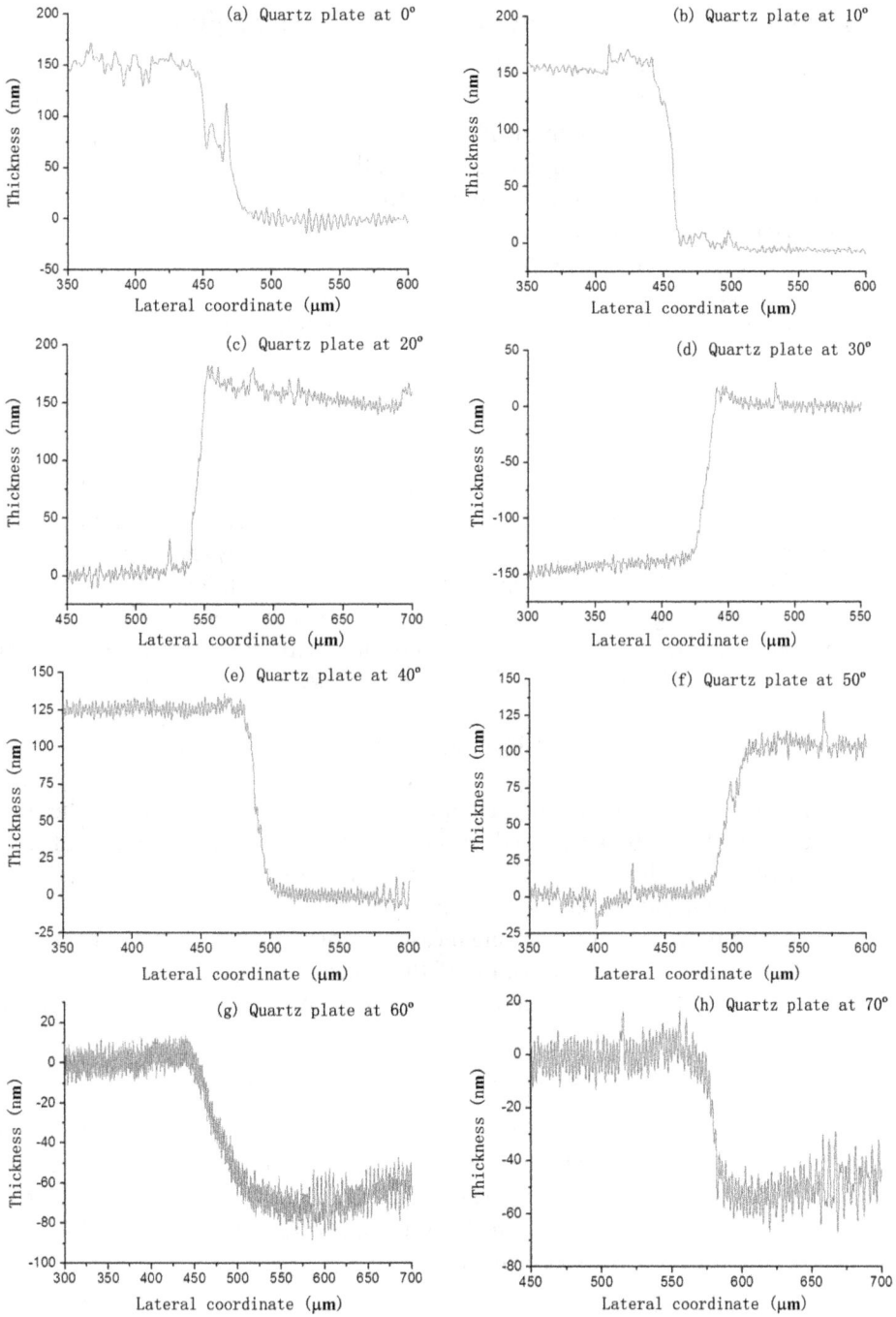

Figure 8.2. Deposit thickness of rhodamine B on different quartz sheets with different inclination angles.

8.4 Error analysis and height correction of the medium layer

Whether the height error introduced by the non-uniformity of layer thickness needs a correction is dependent on the uniformity of the medium deposition and the transfer coefficient of the non-uniformity of layer thickness and height measurement error. As to the confocal scanning measurement, the shape measurement accuracy is decided by the extraction accuracy of the confocal axial peak position. The error is introduced, causing a deviation in shape and height, due to the deviation of confocal axial peak position, caused by the axial convolution between the object function and the medium layer.

The confocal axial response is expressed as the following function when the thickness of the fluorescent medium deposition is t_f on the surface of the sample:

$$I_t(z) = \int_0^{t_f} I_t(z - z')\mathrm{d}z',$$

where $z_0 + t_f > z > z_0$. $I_t(z)$ is supposed to be the confocal axial response of an infinitely thin medium layer deposited on the specimen, while $I_T(z)$ refers to the axial response of a point with relative height errors to be corrected. Since $I_t(z)$ is an even function, by calculating $\partial I_T(z)/\partial z = I_t(z) - I_t(z - t_f)$, it can be seen that the position of the confocal axial peak is at $z = t_f/2$. Therefore, with the thickness (t_f) of the fluorescent medium, the deviation of the confocal measurement result is $t_f/2$ from the actual surface position of the specimen; that is, the transfer coefficient between the thickness error of the medium layer and shape measurement error is 2:1.

According to optical imaging theory, the convolution integral will not only cause axial peak shift, but also broaden the confocal axial response envelope. The axial peak displacement can be calculated by comparing the width difference of axial response of thick and thin medium layers. Eventually the height error of the shape measurement introduced by the thickness non-uniformity of the fluorescent medium layer can be eliminated by compensating measurement results for the axial peak shift.

Supposing the actual height position of the specimen is z_0, the thickness of the medium layer t_f, the object function of the medium layer rect[z], then the confocal axial response is

$$I_T(z, t_f) = I_t(z) \otimes rect\,[(z - z_0 - t_f/2)/t_f].$$

Since $I_t(z)$ is not an analytic function, it is represented as $I_t(z) = \mathrm{sinc}^2(z/a)$ by means of numerical approximation, where a can be confirmed by fitting $I_t(z)$ to the function of $\mathrm{sinc}^2(z/a)$. Then, by fitting $I_t(z) \otimes rect[(z - z_0 - t_f/2)/t_f]$ to $I_T(z, t_f)$, the value of t_f can be determined, and compensation for thickness error can be accomplished by subtracting $t_f/2$ from the measurement result.

8.5 Application of medium aided scattering confocal microscopy

In order to achieve section imaging and shape measurement of smooth, highly curved, or tilted surfaces, a confocal microscope based on medium layer scattering

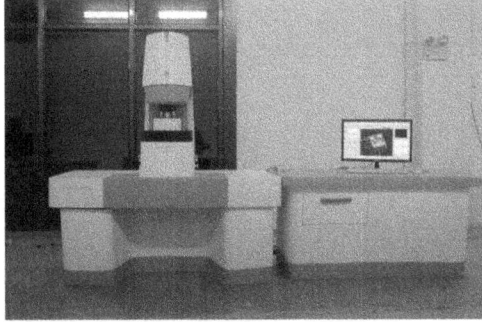

Figure 8.3. Confocal profiler based on medium aided scattering.

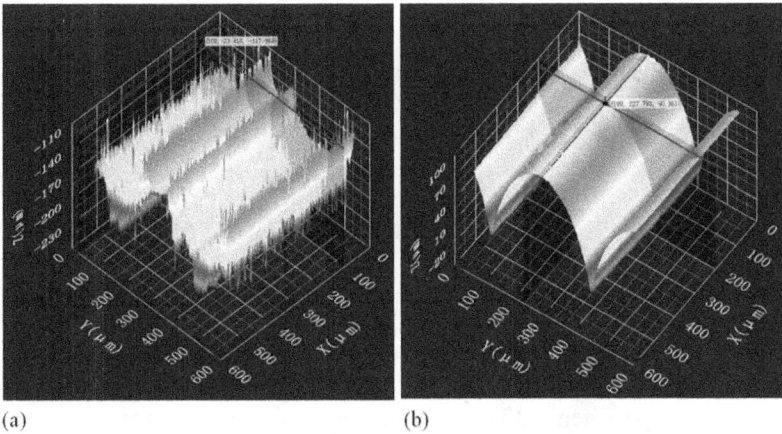

(a)   (b)

Figure 8.4. 3D shape measurement of cylindrical lens array. (a) Result of bright-field confocal. (b) Result of confocal microscope based on medium layer scattering.

has been developed by Harbin Institute of Technology (HIT). An image of the microscopy is shown in figure 8.3.

A cylindrical microlens array with height 120 μm and period 335 μm has been selected as the testing specimen, which is difficult to observe due to its transparency and low reflectivity. The thickness of the rhodamine B medium layer deposited on the surface of the specimen is 100 nm, the numerical aperture of the objective is 0.3 and the wavelength of the laser illumination is 532 nm. The three-dimensional (3D) measurement result of the bright-field confocal is shown in figure 8.4(a) and the result of a confocal microscope based on medium layer scattering is shown in figure 8.4(b).

As shown in figure 8.4, the bright-field confocal microscope cannot resolve the details of the cylindrical microlens array, especially in the region near the junction of the microlenses, while the medium layer scattering confocal microscope offers a complete 3D profile. A cross section view of the cylindrical microlens array is drawn in figure 8.5, with the measurement results $H = 119.5$ μm, $L = 332.9$ μm, and

(a)

(b)

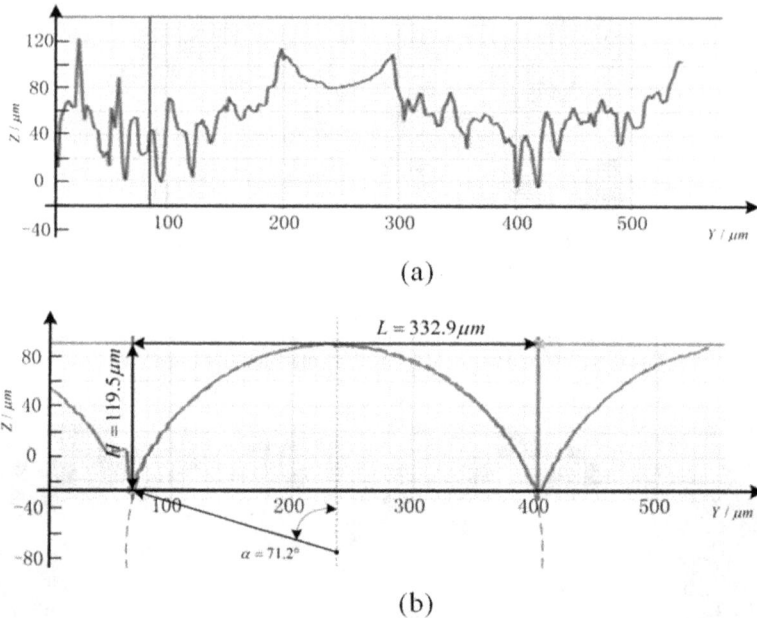

Figure 8.5. Section of cylindrical lenses array. (a) Section of bright-field confocal result. (b) Section of fluorescent confocal result.

maximum slope angle 71.2°. However, with a 0.3 NA objective the theoretical measurable slope angle of a bright-field confocal microscope is only ±18°. The conclusion is that a confocal microscope based on medium layer scattering is easily able to measure a specimen with a highly curved or tilted surface.

References

[1] Avendaño-Alejo M, Moreno-Oliva V I, Campos-García M and Díaz-Uribe R 2009 Quantitative evaluation of an off-axis parabolic mirror by using a tilted null screen *Appl. Opt.* **48** 1008–15

[2] Bartl G, Krystek M, Nicolaus A and Giardini W 2010 Interferometric determination of the topographies of absolute sphere radii using the sphere interferometer of PTB *Meas. Sci. Technol.* **21** 115101

[3] Bayer F L, Hu S, Maier A, Weber T, Anton G, Michel T and Riess C P 2014 Reconstruction of scalar and vectorial components in x-ray dark-field tomography *Proc. Natl Acad. Sci.* **111** 12699–04

[4] Chyan W, Zhang D Y, Lippard S J and Radford R J 2014 Reaction-based fluorescent sensor for investigating mobile Zn^{2+} in mitochondria of healthy versus cancerous prostate cells *Proc. Natl Acad. Sci.* **111** 143–8

[5] Forbes G 2012 Characterizing the shape of freeform optics *Optics Express* **20** 2483–99

[6] Forrest S R 2004 The path to ubiquitous and low-cost organic electronic appliances on plastic *Nature* **428** 911–8

[7] Garnaes J 2011 Diameter measurements of polystyrene particles with atomic force microscopy *Meas. Sci. Technol.* **22** 094001

[8] Guffey M J and Scherer N F 2010 All-optical patterning of Au nanoparticles on surfaces using optical traps *Nano Lett.* **10** 4302–8

[9] Gutin M, Gutin O, Wang X-M and Ehlinger D 2013 Interferometric tomography metrology of conformal optics *SPIE Defense, Security, and Sensing* International Society for Optics and Photonics

Chapter 9

Scanning technology

9.1 Introduction

In confocal microscopy, the introduction of pinholes effectively inhibits stray light and improves a system's signal-to-noise ratio (SNR). However, only one point can be detected each time. The point imaging in confocal microscopy is able to obtain high resolution at the expense of the field of view (FOV), while scanning technology and scanners effectively remedy confocal microscopy imaging defects. Scanning technology and scanners have been well-developed for a long time, and have mainly been concerned with characteristics including a large FOV, high precision, and long device lifetime. Raster scanning is one of the current popular scanning technologies owing to its easy reconstruction of acquired data and good scanning repeatability. Another emerging scanning technology is α-β circular scanning with the advantages of a large FOV and high scanning reliability.

9.2 Scanners

In optical confocal microscopes, scanners are needed to realize the surface measurement of a three-dimensional (3D) structure. There are various scanners available for diverse practical applications, with several factors to be considered including: large FOV, high reliability, and reasonable cost.

Different scanners are commonly used for point scanning imaging, including polygon mirrors, galvanometers, Risley prisms, and microelectromechanical systems (MEMS) scanners, etc. Polygon mirrors feature a large scanning range (the maximum scanning angle is approximately 180°) and high scanning speed (60 krpm). However, the high machining precision requirement of polygon mirrors results in increased manufacturing costs [1]. A well-proven galvanometer must be a versatile laser scanner. In performance, the scanning linearity and speed of a galvanometer can be maintained at 99.9% and kHz level, respectively, the scanning accuracy of a galvanometer can be sustained at μrad level. As micro-optical devices, Risley prisms can attain a speed up to krpm level and their precision can achieve up to 0.5 mrad level. In comparison with a galvanometer, the Risley prism has lower scanning precision and a more complicated mapping algorithm of its data process [2]. MEMS scanners as used in an optical scanning system are compact in design and light in weight, and their scanning speed is up to 3 kHz [3]. However, manufacturing costs of these MEMS scanners is quite high and beam diameter is restricted by the micro-mirror, which causes the incident beam to not take up the full lens pupil size.

Parallel scanners mainly comprise Nipkow [4, 6], micro-lens array [7], digital mirror devices (DMDs) [5, 8], and so on. Parallel scanners are used to split laser beams in order to achieve multi-point illumination and detection, by increasing scanning speed. However, SNR decreases as a result of signal and detector energy segmentation.

As a type of relatively mature scanner, the galvanometer is widely applied in scanning imaging systems, due to its advantages of flexible scanning design, and simple structure, assembly and adjustment.

9.3 Raster scanning

Raster scanning is one of the popular scanning technologies, whose advantages are easy to achieve and it is therefore simple to reconstruct the acquired data.

By inputting different control signals into a galvanometer, various tracks can be obtained easily. Raster scanning is achieved by inputting sawtooth signals and step signals into the fast axis and slow axis of a galvanometer, respectively. Sawtooth signals can control the fast axis to achieve line scanning, and the step signal is used to control the low axis in order to deflect a certain angle and move the scanning spot to the next line. The combination of the fast axis and low axis of the galvanometer achieves raster scanning.

The galvanometer is approximated to a typical second-order low-pass system. From the expansion of the Fourier series, we know that a sawtooth signal comprises high frequency harmonic components, and the amplitude of a high frequency harmonic component increases as the frequency of the sawtooth signal increases. When sawtooth signals are inputted into a galvanometer system, the high harmonic components are attenuated, so that the scanning motion of the galvanometer is not in accordance with the specified track, which results in serious distortions appearing around the boundary areas of the scanning range. We can remove scanning distortion by dealing with acquired data. By raising the frequency of the sawtooth signal input, specifically raising raster scanning speed, the range of distortion will enlarge because more high harmonic components are restrained. This leads to a smaller FOV, if we remove scanning distortion by dealing with acquired data. In addition, the scanning rectangular FOV does not match the circular FOV of an optical system, and therefore is unable to make full use of a valid FOV in said system.

Moreover, for a galvanometer, the main factor that influences its reliability is the impact during scanning motion. The galvanometer is one of the key high frequency devices in confocal microscopy. Raster scanning uses sawtooth signals to control the fast axis to scan. Due to sudden changes in a galvanometer's motion direction at the peak of the sawtooth signal, the acceleration of a galvanometer will increase substantially, resulting in structural impact on the galvanometer. Impact not only influences scanning precision, but also decreases long-term service life.

The sawtooth signal is the fundamental reason why raster scanning is of small FOV and robust reliability. Therefore, α-β circular scanning has been introduced and applied by the Harbin Institute of Technology. The α-β circular scanning FOV corresponds well with the circular FOV of an optical system. Moreover, it uses sinusoidal signals to control the galvanometer, which avoids impact damage to the instrument and improves the system's reliability.

9.4 α-β circular scanning

Section 9.3 mainly introduced the principles and properties of raster scanning. By analysis, the authors can comprehend the fact that sawtooth signals are the fundamental reason why raster scanning is of small FOV and robust reliability. Consequently, a sinusoidal signal is considered and α-β circular scanning is proposed by using sinusoidal signals with a constant phase difference of $\pi/2$ to realize a circular track.

α-β circular scanning FOV matches the effective FOV of an optical system. As shown in figure 9.1, the circular blue area stands for the FOV of α-β circular scanning and the area surrounded by the red line is the FOV of an optical system. The rectangular yellow area represents the FOV of raster scanning. If the relationship between the two FOVs is shown as in figure 9.1(a), 21% of the FOV of raster scanning is ineffective, which not only increases measuring time but also slows down the effective utilization of data. On the other hand, if the relationship between the two FOVs is shown as in figure 9.1(b), that is, on the condition that ineffective

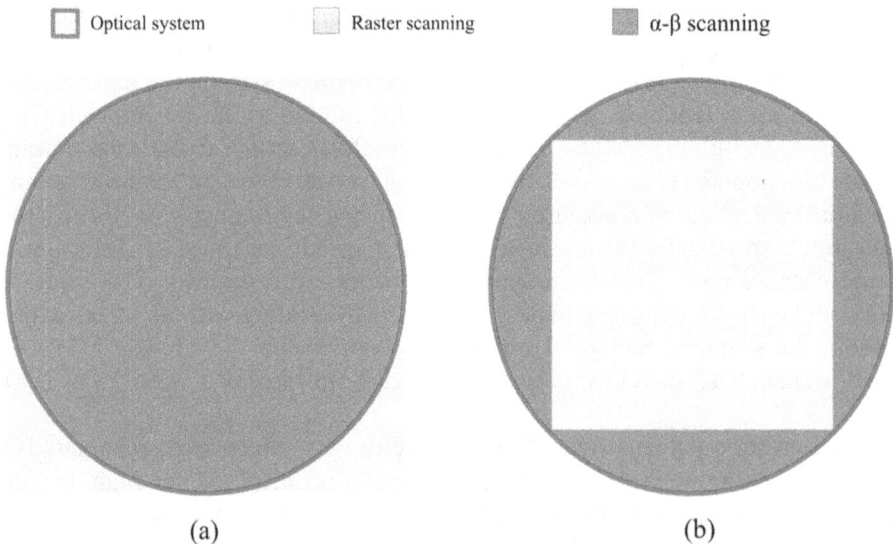

☐ Optical system   ▒ Raster scanning   ▣ α-β scanning

(a)   (b)

Figure 9.1. Comparison of the FOVs between an optical system and raster scanning.

measurement is not produced, the FOV of raster scanning is smaller than the FOV in an optical system and the FOV of the optical system cannot be fully used.

Additionally, as a key component of an optical system, the galvanometer has regularly been used for a long time. Instant impact causes certain damage which leads to a decrease in the scanning accuracy of the galvanometer and shortens its life. Because the impact theoretically depends on the motion direction of the galvan-ometers, the angular acceleration of galvanometers in raster scanning and α-β circular scanning is analy-zed to evaluate the reliability of a galvanometer. The angular acceleration can be obtained by dealing with the second-order derivation of sawtooth and sinusoidal signals. As shown in figure 9.2, at the beginning and the peak of the sawtooth signal of each cycle, the angular acceleration of a galvanometer will suddenly increase because the motion direction of the galvanometer immedi-ately changes. The acceleration increases from $0°$ s^{-2} to $3555°$ s^{-2}. Immediate gains of acceleration at the starting position and at the peak result in great impact on the galvanometer. As the frequency of the sawtooth signal increases, the motion direction needs to be changed in a shorter timespan; therefore, the impact amplitude can be much greater. In α-β circular scanning, the input signal is a sinusoidal one, and the acceleration decreases from $40°$ s^{-2} to $-40°$ s^{-2}. The maximum dynamic amplitude of acceleration is only 1/44 of that in raster scanning. In comparison with raster scanning, α-β circular scanning can effectively weaken instant impact and damage to a galvanometer, elongating the lifetime of the galvanometer and improving the reliability of the galvanometer.

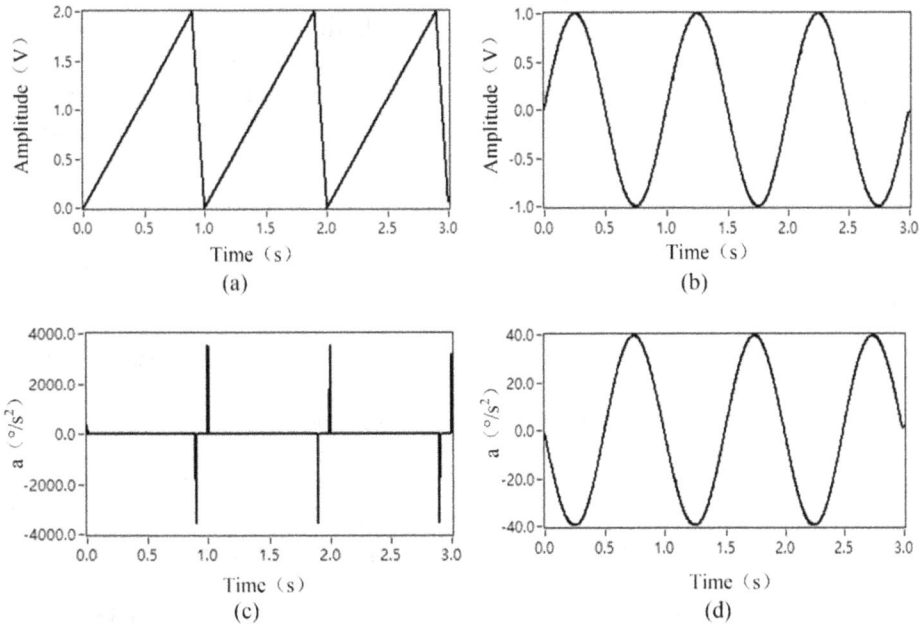

Figure 9.2. First and second derivatives of sawtooth and sinusoidal signals. (a) 1 Hz sawtooth signal. (b) 1 Hz sinusoidal signal. (c) Angular acceleration, the second-order derivative of the sawtooth signal. (d) Angular acceleration, the second-order derivative of the sinusoidal signal.

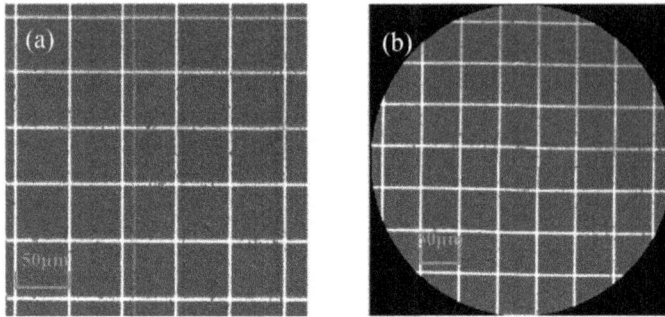

Figure 9.3. Scanned images. (a) Raster scanning, (b) α-β circular scanning.

In addition, experiments have been conducted to validate the fact that α-β circular scanning has the advantage of a large FOV. A grid sample is measured by using two scanning techniques. Raster scanning images are shown in figure 9.3(a). If raster scanning images are not distorted, the FOV is 75625 μm^2. Figure 9.3(b) is α-β circular scanning imaging, and the FOV is 119459 μm^2. By the above experimental analysis, the conclusion can be drawn that the FOV of α-β circular scanning is enlarged by 57% in comparison with raster scanning.

This chapter mainly introduces commonly used scanning technology and scanners in confocal microscopy. Scanning technology and scanners resolve the conflict between a large FOV and high resolution. Galvanometers with the advantages of flexible scanning track, easy installation, and simple operation are widely applied in confocal microscopy, which can realize raster scanning and α-β circular scanning. Raster scanning is regarded as acceptable due to its easy reconstruction and good repeatability. α-β circular scanning, as a new scanning technology, has the advantages of a large scanning FOV and high scanning reliability, which shows promising potential in scanning imaging, optical tweezers, laser beam fabrication, etc.

References

[1] Duma V-F and Rolland J P 2014 Advancements on galvanometer scanners for high-end applications *J. Proc. SPIE* **8936** 893612–2

[2] Warger W C and DiMarzio C A 2007 Dual-wedge scanning confocal reflectance microscope *Opt. Lett.* **32** 2140–2

[3] Ra H, Piyawattanametha W, Taguchi Y, Lee D, Mandella M J and Solgaard O 2007 Two-dimensional MEMS scanner for dual-axes confocal microscopy *J. M.S* **16**

[4] Tanaami T *et al* 2002 High-speed 1-frame/ms scanning confocal microscope with a microlens and Nipkow disks *J. Appl. Opt.* **41** 4704

[5] De Beule P A A, deVries A H B, Arndt-Jovin D J and Jovin T M 2011 Generation-3 programmable array microscope (PAM) with digital micro-mirror device (DMD) *C. SPIE* **7932** 79320G–1

[6] Grant D M *et al* 2007 High speed optically sectioned fluorescene lifetime imaging permits study of live cell signaling events *J. Opt. Express* **15** 15656–73

[7] Tiziani H J and Uhde H M 1994 Three-dimensional analysis of confocal microscopy with microlenses *J. Appl. Opt.* **33** 567–72

[8] Hou W and Zhang Y 2011 Fast parallel 3D profilometer with DMD technology *C. SPIE.* **8321** 1–7

Chapter 10

Confocal profilometer

10.1 Introduction

A confocal profilometer can be used for the profile measurement of optical elements [1,2]. It is an optical non-contact measurement instrument with the characteristic of high-precision [3–6]. It also has the functions of macro-profile measurement and microstructure detection. Therefore, it is suitable for the profile measurement of large diameter hybrid aspherical diffractive infrared elements (HADIEs), which have the characteristics of the macro–micro combination [5]. Moreover, confocal profilometers can provide closed loop detection to improve fabrication accuracy for optical element manufacture.

10.2 Basic principle

A confocal profilometer uses x-direction and z-direction associated motion models to achieve the functions of geometrical profile and local area three-dimensional measurement, in which a high-precision grating ruler is used as a position feedback sensor. The x-direction movement unit consists of an airborne platform, driving motor, high-precision grating ruler, and other major devices and components. The purpose of using a precision airborne platform is to reduce axial and radial flow rates, thereby reducing errors of measurement. The z-direction scanning unit uses macro–micro positioning to achieve axial movement and local tomography scanning, and consists of manual adjustment mechanisms and a high frequency micro-driver. In the process of profile measurement, the x–y scanning mechanism of confocal remains static, and the system only needs tomography scanning by the z-direction scanning unit. When microscopic observation is required, the optical x–y scanning mechanism and z-direction tomography scanning mechanism, cooperate to achieve three-dimensional microscopic observation and measurement of a local region. An image of a confocal profilometer is shown in figure 10.1.

doi:10.1088/978-1-6817-4337-0ch10   © Morgan & Claypool Publishers 2016

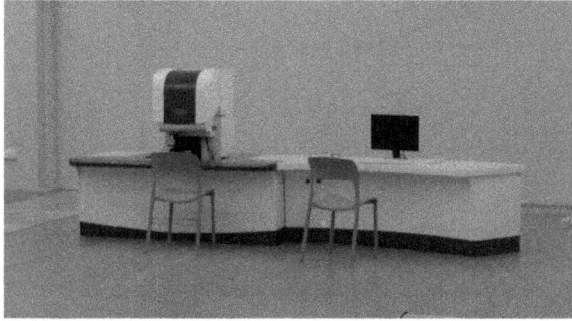

Figure 10.1. Photo of confocal profilometer.

10.3 The extraction method of discrete surface

HADIEs are representative optical elements with microstructures based on curved surfaces. The difficulty in profile measurement of the elements, is that macro-profile fitting is seriously influenced by the extraction precision of micro-step height. A special fitting method for the measurement instrument is needed to solve the problem of continuity of the step edges based on a curved surface. Therefore, the surface profile equation of the sample can be fitted precisely. To calculate the height of the step based on the curved surface, planar substrate conversion and polynomial regression methods have been proposed.

10.3.1 Method of planar substrate conversion

For the evaluation of a step height of a known profile, the equation can be converted to steps on a plane in order to calculate according to theoretical equations. HADIEs are combined with the profiles of aspherical and diffractive elements. They are considered as a combination of continuous aspherical profiles and vertical steps. The equations of the continuous aspherical and step are

$$z = \frac{cx^2}{1 + \sqrt{1 - (k + 1)c^2x^2}} + \sum A_{2n+2}x^{2n+2} + \frac{1}{n_0 - 1}\sum H_{2n}x^{2n} \tag{10.1}$$

$$z = \frac{\lambda_0}{n_0 - 1}\left| Int\frac{1}{\lambda_0}\sum H_{2n}x^{2n}\right|. \tag{10.2}$$

Steps can be obtained by subtracting the designed value from the measured value after tilt and shift corrections. When the fabrication errors are very small, step height can be calculated using the single side step calculation principle. To calculate step height, we can use the least square curve fitting polynomial

$$z = a_0 + a_1x + h \cdot \delta. \tag{10.3}$$

The core concept of the method of planar substrate conversion is that it can be converted to a step on a plane from a curved base, thus it can improve the actuality

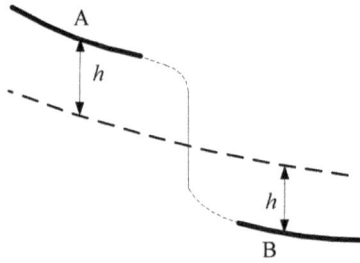

Figure 10.2. The step based on a curved surface.

and stability of the calculation of step height. The necessary condition is that the profile formula is known.

10.3.2 Method of polynomial regression

According to the mathematical model of HADIEs, the upper and lower surfaces of each step correspond to the same function (in theory) as shown in figure 10.2. Therefore, the two parts of measuring data A and B can be calculated using the least square curve fitting polynomial regression method

$$z = a_0 + a_1 x + a_2 x^2 + \cdots + a_n x^n + h \cdot \delta, \tag{10.4}$$

where $a_0, a_1, a_2 \ldots a_n$, and h are undetermined parameters, and $\delta = \pm 1$ differentiates between the top and bottom curves. The height d of the step is two times that of the estimated h value.

The polynomial regression method is not affected by the precision of measurement positions of step edges, and can also increase a constraint condition of the polynomial fitting at the edge of steps. It is suitable for the calculation of step height based on a continuous curved surface. The precision of polynomial fitting also depends on the selection of polynomial order. Generally speaking, the use of a higher order polynomial can reach a higher fitting precision. It cannot always increase the precision of data processing effectively if the polynomial order passes some certain value. The simulated results show that the best value of the polynomial order is 2.

10.4 Application of confocal profilometer

The mathematical expression for the profile of the HADIE as processed is shown as follows:

$$z = \frac{cx^2}{1 + \sqrt{1 - (k+1)c^2 x^2}} + Ax^4 + Bx^6 + P_1 x^2 + P_2 x^4, \tag{10.5}$$

where $c = 0.00534559$, $k = 4.2335$, $A = -4.529 \times 10^{-8}$, $B = -4.2966 \times 10^{-12}$, $P_1 = -1.11511383 \times 10^{-5}$, $P_2 = -2.6693989 \times 10^{-9}$, and the step height is 3.66 μm.

The first manufactured HADIE's measurement results for geometric profile are shown in figure 10.3.

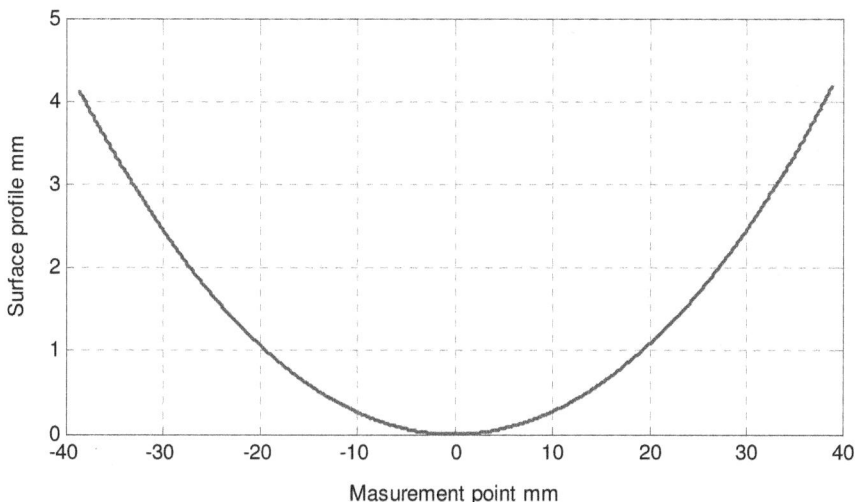

Figure 10.3. The original measurement data of a HADIE.

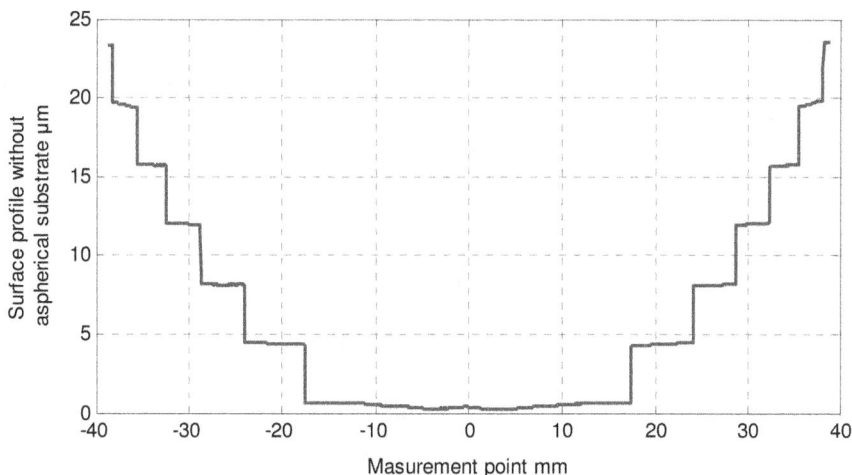

Figure 10.4. Rest steps after removing the aspherical profile.

The diameter of the measured HADIE is 80 mm, the difference in height is about 4.5 mm, and the theoretical value of steps is 3.66 μm. As a result, the original measurement data profile is approximately an aspherical profile. After removal of the aspherical profile, the profile of the steps can be clearly seen. The result is shown in figure 10.4.

Firstly, we select the effective data from the original measurement data and correct tilt and translation. Then we calculate the step height, eliminate the steps, obtain a continuous aspherical profile data, and process this data by a Gaussian filter. Lastly, from the aspherical profile data, we subtract the designed profile data of the element to obtain the errors of HADIE after first fabrication by diamond lathe. The PV value is about 10.5 μm as shown in figure 10.5.

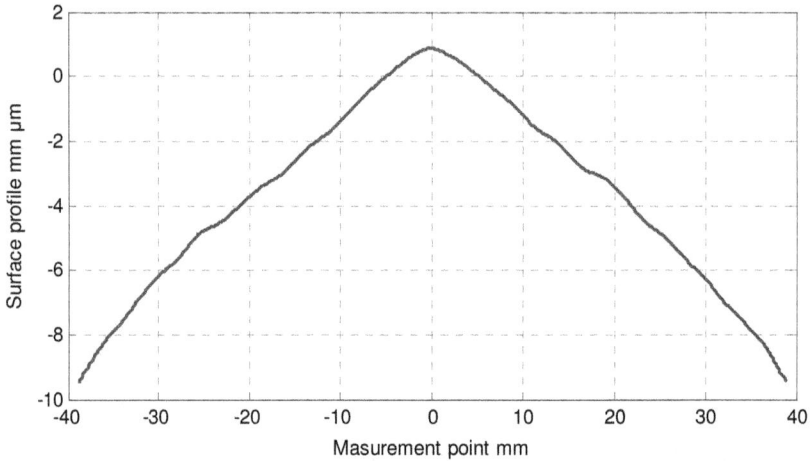

Figure 10.5. The error in the first fabrication of a HADIE.

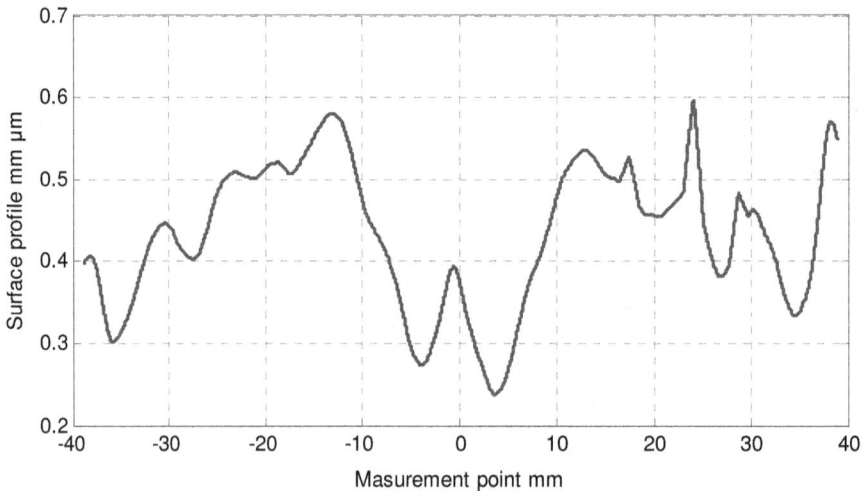

Figure 10.6. The profile error of a HADIE after compensation.

We translate the first fabrication error of a HADIE to a compensation file that can be recognized by a diamond lathe. The file guides the compensation fabrication of the HADIE. We measure the HADIE, which has undergone a compensation process, to obtain the profile errors after data processing. As shown in figure 10.6, after the compensation fabrication and adjustment of c value, the result is that the PV value of the profile error is reduced to less than 0.4 μm.

A confocal profilometer can function as a non-destructive tool and be used to achieve the high-precision measurement of HADIEs. Most importantly, it can be used to compensate for fabrication error, particularly in the case of fine optical elements. Moreover, the confocal profilometer can provide three-dimensional measurement to a microstructure with large optical elements and detect the processing defects of the microstructure.

References

[1] Rayer M and Manseld D 2014 Chromatic confocal microscope using hybrid aspheric diffractive lenses *Proc. SPIE* **9130** 91300Z

[2] Conchello J 1997 Novel reflected light confocal profilometer. *Proceedings of SPIE - The International Society for Optical Engineering* **2984** 101–112

[3] Nouira H, Salgado J-A, El-Hayek N, Ducourtieux S, Delvallée A and Anwer N 2014 Setup of a high-precision profilometer and comparison of tactile and optical measurements of standards *Meas. Sci. Technol.* **25** 044016

[4] Tien A 2014 High speed confocal 3d profilometer: design, development, experimental results Ph D.

[5] Liu J, Wang Y, Gu K, You X, Zhang M, Li M, Wang W and Tan J 2016 Measuring profile of large hybrid aspherical diffractive infrared elements using confocal profilometer *Meas. Sci. Technol.* **27** 125011

[6] Tan J, Liu C, Liu J and Wang H 2016 Sinc^2 fitting for height extraction in confocal scanning *Meas. Sci. Technol.* **27** 025006

www.ingramcontent.com/pod-product-compliance
Lightning Source LLC
Chambersburg PA
CBHW082109210326
41599CB00033B/6649

* 9 7 8 1 6 8 1 7 4 3 3 6 3 *